SATELLITE TELEVISION
A Layman's Guide

Dedicated to

My wife Joyce, for her encouragement and tolerance throughout the compilation and production of this book.

SATELLITE TELEVISION

A Layman's Guide

Peter S. Pearson

ARGUS BOOKS

Argus Books Limited
1 Golden Square
London W1R 3AB

ISBN 0 85242 906 1

Phototypesetting by En to En, Tunbridge Wells
Printed and bound by Whitstable Litho

Contents

Acknowledgements

The author would like to thank the following, for providing information and some of the photographs.

ATN/Filmnet, Amsterdam
J. Battye
BBC
S. J. Birkhill
Galaxy Satellite Television
Huddersfield Examiner
Satellite TV Antenna Systems Ltd (SATVRN)
Satellite TV Publications Ltd
Satellite World, USA

Introduction 1

Satellite television is a phrase being used quite frequently these days around Europe. What is it, and how is it likely to influence your viewing habits? This book is written in layman's language, and is intended to provide the answers to many questions relating to the installation and operation of a Television Receive Only (TVRO) satellite terminal.

In order to understand what satellite television is all about, it is necessary to explain something about the history of this technological development, and this really stems from an article in a technical magazine *Wireless World* in the mid-1940s by Arthur C. Clarke, the man of science fiction who wrote such books as *Childhood's End, 2001–A Space Odyssey*, and more recently the follow up *2010 — Odyssey Two.* Arthur Clarke is an Englishman, who for the past thirty years or so has lived in Sri Lanka, where he concentrates on his writings and his work at the Arthur Clarke Centre, which is part of the University of Moratuwa.

The article in *Wireless World* in 1945 expressed Clarke's idea that if it were possible to place satellites in orbit above the equator, at a precise distance, approximately 22,300 miles above the Earth, then the orbital speed of the satellites would match the speed of rotation of the Earth, and so from any point on Earth these satellites would appear to be stationary. They are said to be in a 'geostationary orbit'. At the time the idea seemed to be pure science-fiction, and not many people took it seriously.

Later, when powerful rockets were being developed, and space exploration became a possibility, engineers started to think again about placing satellites in orbit. Electronics research and development engineers started to design communications equipment suitable for operation in satellites and Arthur Clarke's ideas suddenly became practicable. Although many other scientists have contributed to this development, it is a nice thought that the precise area where all the communication satellites now orbit the Earth is called 'The Clarke

Belt', a tribute to the man who was the first to publish the idea of space communication by satellite.

At this point it might be worth mentioning why Arthur Clarke thought it would be a good idea to transmit radio and television programmes from space. The economics side of the coin is the fact that it requires a large number of transmitting stations on Earth to communicate with the population. For example, the BBC and IBA need hundreds of such stations just to cover the United Kingdom. When you consider that one satellite could not only cover the whole of the UK, but most of Europe as well, then it is obvious that the economics make sense. There are other considerations, of course, but the basic concept is a good one.

The first few communications satellites that were launched could not be put into geostationary orbit because the rocket techniques had not then been developed. The USSR launched the very first artificial satellite *Sputnik* in 1957, and it was five years before the USA launched its first artificial satellite *Telstar*. The first satellite for television relay, *Telstar*, was not geostationary and had to be tracked by earth stations as it passed overhead a few times every day. It was 1965 before the USA launched the first geostationary communications satellite *Early Bird*, which could relay several hundred telephone channels, or one television channel. In the next 18 years or so, the USA and Canada put another 15 geostationary communications satellites in orbit. In mid-1986, when this book was being compiled, there were over 100 communications satellites in geostationary orbit, and another 100 plus are already planned (Figure 1.1).

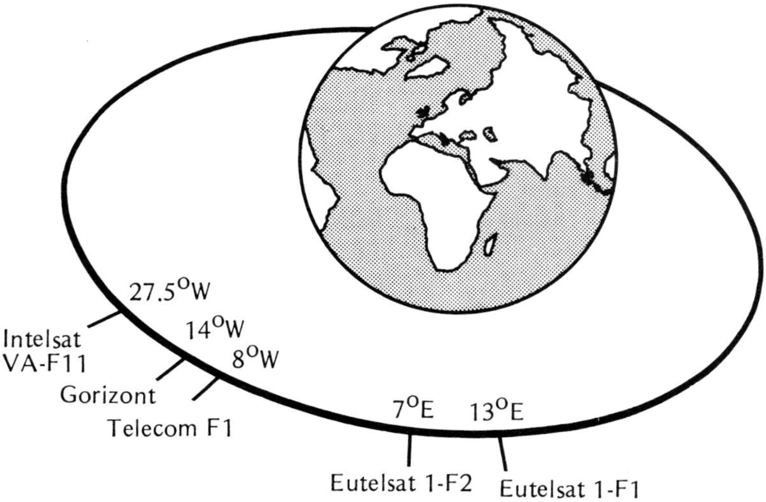

Figure 1.1 Pictorial representation of satellites in geostationary orbit.

There are two principal methods of placing artificial satellites into orbit. One is to use an unmanned rocket to take the satellite from Earth and put it into geostationary orbit. This method is used by the USSR, China and the European Space Agency in France which uses the Ariane rocket. The other method, favoured by the USA, is the use of the manned Space Shuttle. However, at the time of writing, the future of the space shuttle is under review following the disastrous explosion which wrecked the Challenger Shuttle whilst it was being launched in January 1986.

Every satellite carries a certain amount of fuel to power small booster rockets, and these can be remotely controlled from Earth, as required to make certain alterations in the position of the satellite. The amount of fuel carried is estimated so that corrections can be carried out for the entire life of the satellite, which can be anything from 2 or 3 years up to about 10 years, although the majority are intended to have a useful operating life of between 3 and 5 years. At the end of its life a satellite can be moved out of its station and a new satellite put in exactly the same position. The old satellite then becomes just another piece of 'space junk', joining the thousands of spent rockets and other pieces of debris which are now floating about in space.

Each communications satellite carries a number of *transponders*. These are the devices which actually deal with the signals. Normally each transponder carries one programme channel. The power to operate the electronics comes from a series of solar panels which convert light energy picked up from the Sun, into electrical energy to power the equipment. There are usually a number of transponders on each satellite and each of these can carry either a separate television programme, a large number of telephone circuits, radio programmes or digital data such as is used in computer links, etc. As the amount of power available is obviously limited there are one or two arrangements which are employed by the designers in order to use the power available in the most economical way, such as using half a transponder for one signal and the other half for something else.

The satellites discussed in this book should not be confused with the Direct Broadcasting by Satellite (DBS) system, which has been the subject of much debate over the last few years. The whole subject is now being reconsidered, and in the meantime the industy is pushing ahead with the development of equipment to utilise the low power signals, originally intended for the use of cable companies, which have now proved to be a viable proposition for home reception, using dishes not so much larger than the one metre dish proposed for the DBS system.

The satellites we are concerned with were not intended originally for direct reception in the home. It is only because one or two enthusiastic amateurs set about to receive the very low-level signals, by developing sophisticated electronics, that home TVROs are possible today.

The very first home TVRO terminal in the world was built by a Yorkshireman, Steve Birkill, in 1975, at his home near Sheffield in England. At that time Steve was a senior engineer with the BBC Transmitter Department. Since his schooldays he had been interested in receiving television over long distances. He was a radio ham and developed many new ideas for long-distance TV. His parents' home in Barnsley was distinguishable by the high scaffold pole above the roof which carried large aerials that could be turned to face the transmitters. Steve was a world leader in this field, so when satellite TV was evolved it was a logical development of Steve's hobby to try to receive it. As so often happens, the hobby became the most important aspect of his life, and in the early 1980s Steve decided to leave the BBC and concentrate on satellite TV. Today he is the Technical Director of one of the very few British firms manufacturing TVRO equipment. He is responsible for the research and development of equipment, and in addition he is the Europe Bureau Chief for the European edition of the American magazine *Satellite World,* which is a monthly publication devoted to satellite TV.

The second home TVRO terminal was built by the very man who first published the idea for satellites in geostationary orbit, and this was, of course, Arthur C. Clarke whose terminal is in Sri Lanka. The first home terminal in America was built nearly three years after the Sheffield one by Robert 'Bob' Cooper at his home in Oklahoma. Bob has collected a vast amount of information relating to satellite TV, and has published several books and leaflets about it.

In less than a decade, the home TVRO industry in the USA has become probably the fastest-growing business in the world, and now, with the British Government's relaxation of the rules governing reception of satellite TV, it seems likely that it will soon become a booming industry in Europe, and the UK in particular.

Transmission 2

Television programmes, and radio programmes for that matter, normally originate in studios probably many miles from the listener and viewer. From the studios the signals are sent along telephone lines or radio links to several transmitting stations, situated in different parts of the country. These transmitting stations are frequently situated on high ground and can be recognised by their tall masts or towers which support the aerials (Figure 2.1). These masts and towers are often referred to by the press, and indeed sometimes by the BBC and commercial TV companies themselves, as transmitters. This is not an accurate statement because the transmitters are actually cabinets of electronics which are housed in the buildings close to the mast base. The masts are simply the supports for the aerials which pump out the energy from the transmitters into the surrounding atmosphere. The small aerials which are fixed to your chimney, or are in the loft, pick up small amounts of this transmitted energy, convey it to the household TV or radio set, and then circuits inside the cabinets reproduce a fair copy of the original sound and pictures that originated in the studios.

Each terrestrial transmitting station has a limited area where its signals can be received. This is determined by the local terrain. Hills and tall buildings reflect the signals, and this causes certain problems such as multiple images (ghosts) on television pictures and certain kinds of distortion on FM radio broadcasts. The amount of energy required by these transmitting stations, the number required to serve the majority of the population, together with the number of staff required to maintain the service, make the provision of radio and television by this method a very expensive business.

In the USA, instead of building a large number of transmitting stations so that people could receive the programmes 'off-air', they instead install cable networks to provide numerous programmes simultaneously. The cable companies receive programmes by landline or radio link from the studios and then pipe these around their area.

Figure 2.1 Television transmitting station at Emley Moor, near Huddersfield, England.

Revenue for this service is collected by the cable company, and the rental is varied according to the number of servcices a client uses. This system is very popular and a large number of cable networks have been established in all the populated areas of the USA and Canada.

However, renting cables is expensive, so an alternative was sought. This leads us to the idea of using artificial satellites for television and radio distribution.

Think about the terrestrial transmitting station with its tall mast to support the aerials. Perhaps the mast is 1000 feet (330 metres) high, and the area it serves is perhaps only 50 miles radius, and this probably has many blind spots which require additional relay stations. What then about a mast 23,000 miles high. It does not take a lot of imagination to realise that the area covered from such a height is enormous. In fact a satellite in the Clarke belt can see about 40% of the Earth's surface.

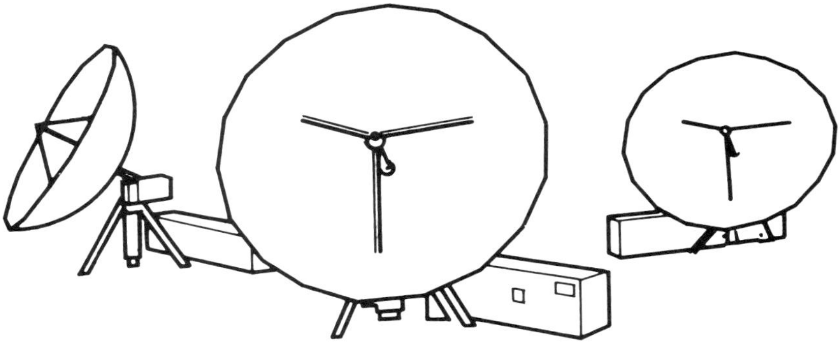

Figure 2.2

Now let us take a look at how the signals get from the studios to the satellite and back again to Earth. The frequencies which are used for satellite communications are much higher than the ones used by terrestrial transmitters. This means the signals have a much shorter wavelength. A very narrow beam is required, both to get the signals to the satellite (the *uplink*), and back again to Earth (the *downlink*). In order to produce this narrow beam a parabolic dish is used, and various companies, usually government bodies responsible for telecommunications, like British Telecom in the UK, and the PTT in France and Belgium, are responsible for providing the uplinks. The signals from the various programme providers are sent to the uplink terminals, in a similar way to which they are distributed to the terrestrial transmitter network, and then these signals are sent to the particular satellite by carefully pointing the parabolic dish at the 'bird'. A dish can only access one 'bird' at any one time, so if you see one of these groundstations it will probably have several dishes, each one pointing at a different satellite (Figure 2.2). Although each dish can only point at one satellite at a time, as mentioned earlier each satellite carries a number of transponders, and each of these transponders is tuned to accept a particular frequency, so one dish can in fact access several transponders on a particular satellite.

A transponder is really a receiver and transmitter. A parabolic aerial on the satellite picks up the signals sent from Earth and routes these signals to the transponders. Each transponder then re-transmits the signals, having changed the frequency, to another parabolic dish pointing back to Earth. The signals from the transponders cover an area on Earth called a *footprint* (Figure 2.3). This footprint is a series of contours, where the strongest signal is at the centre and the signals set progressively weaker as the contours get larger. Dishes within the footprint can be pointed at the satellite and the signals received can be processed and fed into your television or radio receiver. As the power available, on the satellite, is limited, the equipment designers arrange the footprint of particular signals to fall only in an area where they can be useful. For this reason, many footprints are arranged to be strong on land but fall away sharply over large sea areas. At this stage it might be worth while to point out that satellite TV is *not* world-wide TV. It is only possible to receive the signals which fall within the footprint of a particular transponder.

Having said all this, the footprints always cover a much larger area than a terrestrial transmitter, and it is because of this that people in the UK, for instance, can now take advantage of such programmes as Moscow 1 domestic programme, Italy RAI Uno, the equivalent of BBC1 in Italy, together with many other programmes in English, Dutch and German (Figure 2.4). These will be dealt with in more detail later in the book.

So it was in the 1970s that the North American cable companies decided to use satellites for their programme contributions, thus allowing their customers many more programme channels to choose from, and reduce the cost of receiving contributions because of high rental charges for the links. The cable companies in the United States and Canada have continued to expand, and they now have interactive programming, where their customers can communicate with the cable company office. They have various teletext services, and in some cases these are interactive. Clients can communicate with their local supermarket and order provisions, or even check their bank accounts or arrange payment of bills.

Also in the late 1970s a number of governments, including the British, decided that they should investigate the possibility of using

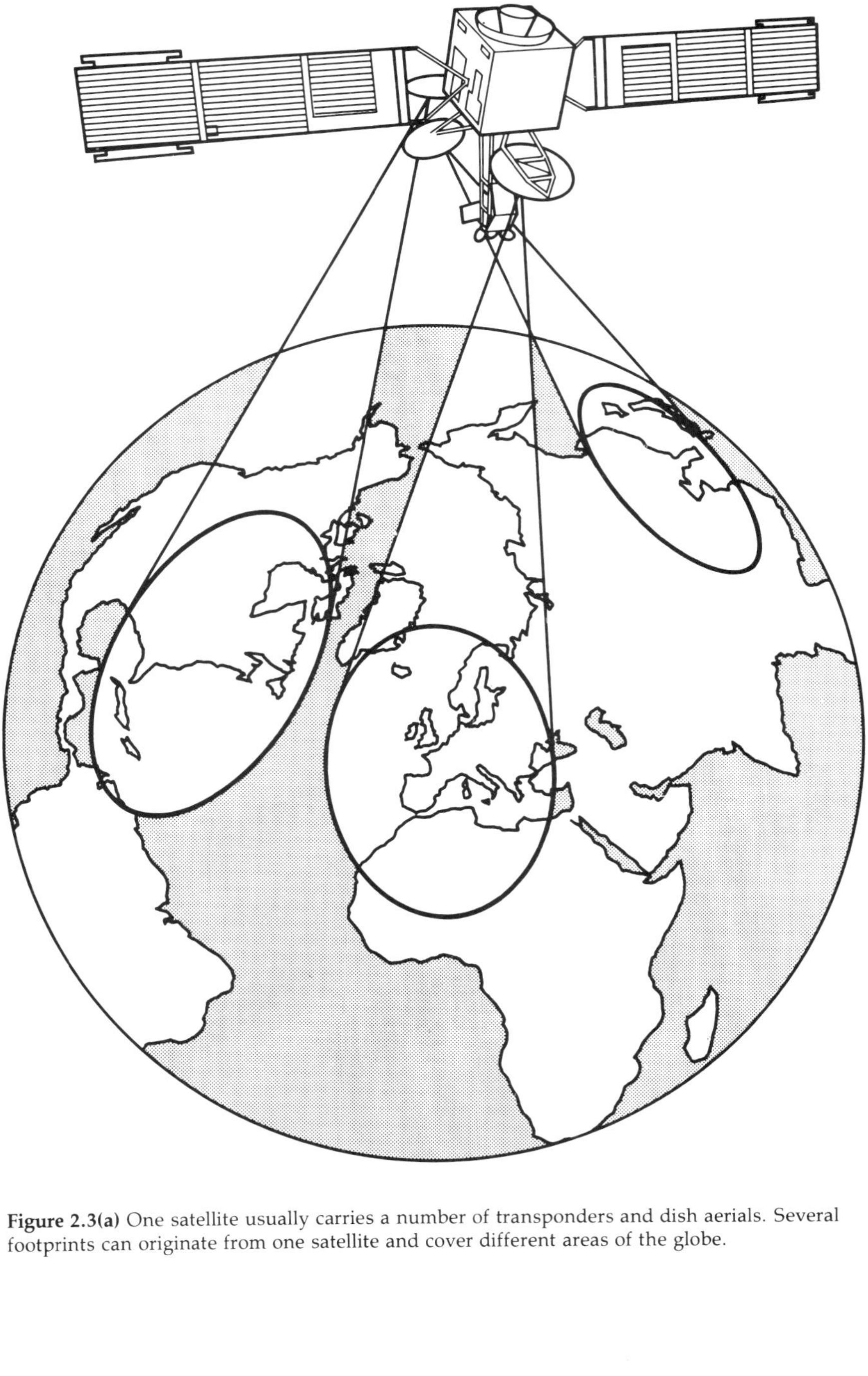

Figure 2.3(a) One satellite usually carries a number of transponders and dish aerials. Several footprints can originate from one satellite and cover different areas of the globe.

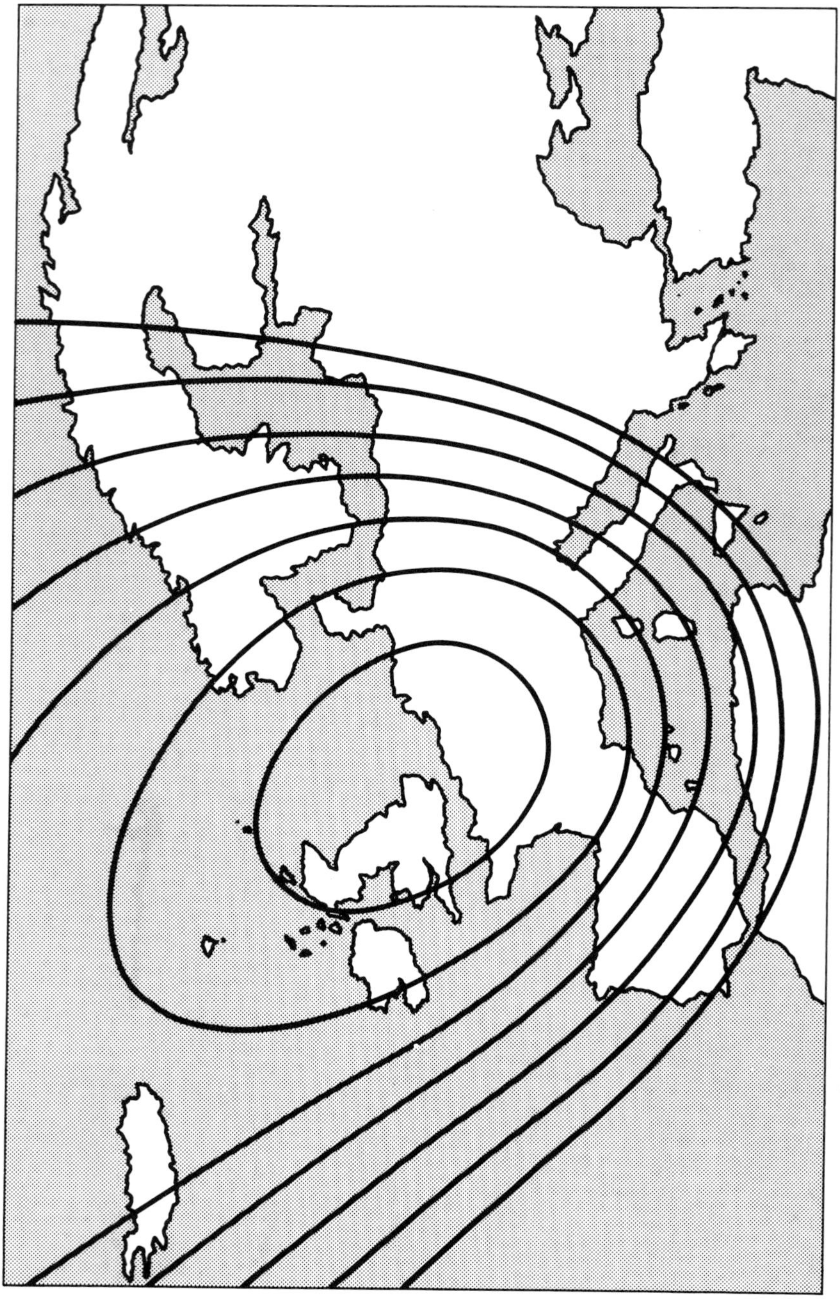

Figure 2.3(b) A footprint of Eutelsat 1 – F1 at 13 degrees East.

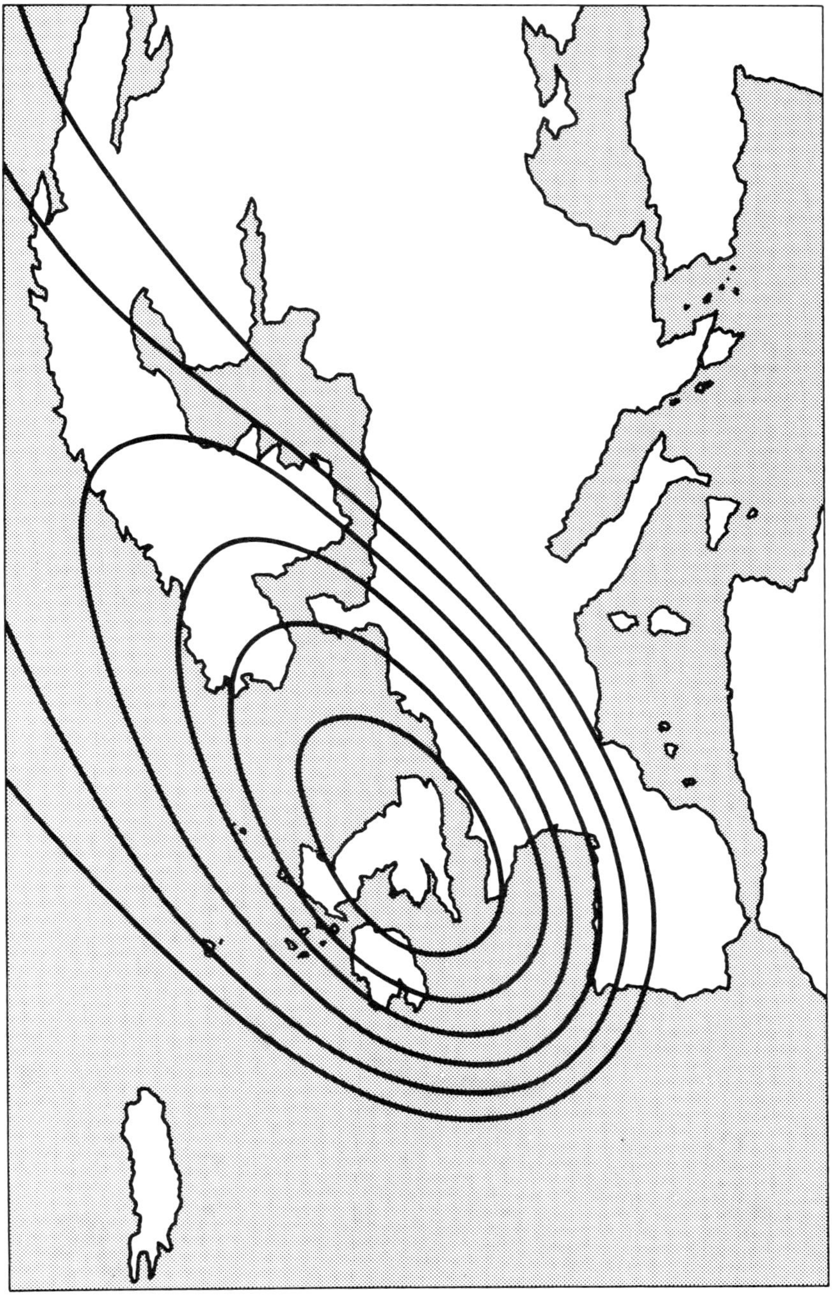

Figure 2.3(c) A footprint of Intelsat VA–F11 at 27.5 degrees West.

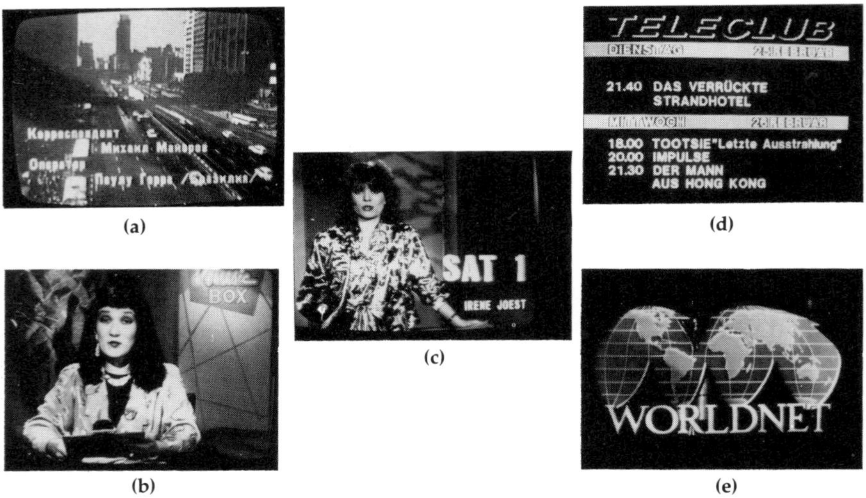

Figure 2.4 Off-screen photographs from satellite programmes: **(a)** TASS statement (Moscow); **(b)** Presenter (Music Box); **(c)** Presenter (SAT 1); **(d)** Programme menu (Teleclub); **(e)** Logo (Worldnet).

satellites for TV. Not only did they consider a similar arrangement to America, whereby cable companies could use this method of programme distribution, but they also considered the possibility of using satellites for TV distribution direct to the homes of viewers. This was to be known as Direct Broadcasting by Satellite (DBS) and, in Britain, a committee was set up, given a number of strict parameters to observe, and told to investigate such a system. One main idea for DBS was that the satellites proposed would have a much higher output power than those used by the Americans for their cable companies, and therefore the signals could be received on much smaller dishes. A size of about 1 metre was anticipated, and this could be fitted easily on most house walls or roofs.

Members of this committee, including the BBC and IBA, deliberated for several years. In the meantime various obstacles emerged. Costs of putting satellites into orbit escalated, and the higher powers stipulated were difficult to achieve on time. In parallel with this, more sensitive receiving equipment was developed and it was discovered that acceptable pictures were receivable from the low-power satellites with dishes as small as 1.2 metres near the centre of the footprint. With a dish of 1.8 metres it was possible to obtain excellent pictures from numerous satellites over a large part of Europe and the UK (Figure 2.5).

The DBS consortium was finally abandoned in 1984/5 and the idea of direct broadcasting by satellite for the United Kingdom was put on ice, pending a re-think. Also in 1985, the British Government relaxed the rule about satellite reception by individuals, TVRO (Television

Figure 2.5 A 1.8 m dish in the garden of a house with a TVRO system.

Receive Only) terminals and small distribution systems, such as hotels and holiday complexes, known as SMATV systems (Small Master Antenna Television). The Home Office issue licences for these at a once-only cost of £10.

In 1986 the question of eventual DBS systems is still being discussed. In fact, some people think that it may even become a reality as soon as 1989. In my opinion, it is much more likely to be at least 10 years before a viable system is in operation. It is probably worth pointing out, however, that any current 11 GHz receiving terminal will be capable of receiving DBS signals, although of course the dish will have to be pointed at the appropriate bird.

At this stage, I should point out that the advent of DBS will probably mean the adoption of a different system of colour transmission. In order to overcome some of the problems encountered with PAL, NTSC and SECAM, it is proposed that DBS will use a system called MAC. This system does not suffer from colour fringing, which is evident on other systems and it lends itself to other improvements, such as higher definition pictures. Possibly it will have a different aspect ratio so that the pictures are more compatible with wide-screen cinema films. In order to receive MAC signals it is necessary to have either a new TV set, or a rather expensive decoder for your existing receiver. The MAC system is at present in use by Norway on Eutelsat 1–F4. I have seen pictures from Norway, and in my opinion it requires a trained engineer to really see the differences, and I wonder if the average viewer will be prepared to buy the expensive decoders for such a marginal improvement.

Some of the most annoying problems, as far as terrestrial viewers are concerned, are due to the hilly terrain causing 'ghosting', and interference from other terrestrial transmissions which produce 'patterning', particularly evident during periods of high sun-spot activity. These problems are not encountered with satellite transmissions anyway, so satellite pictures are generally better than terrestrial ones.

In order to understand the relationship between transmissions, Table 2.1 illustrates their positions in the frequency spectrum

Table 2.1 The frequency spectrum showing the relationship between different transmissions

Band	*Frequency*
Low frequency (LF) Long wave radio (2000–1053 metres)	150–285 kHz
Medium frequency (MF) Medium wave radio (571–187 metres)	525–1605 kHz
Band II FM radio (VHF)	88–108 MHz
Air band	110–145 MHz
Band IV Terrestrial TV (UHF)	470–582 MHz
Band V Terrestrial TV (UHF)	614–854 MHz
Satellite (C-band)	3.7–4.2 GHz
Satellite (Ku-band)	11.7–12.7 GHz
Infrared, visible light, ultraviolet, gamma rays	above 30 GHz

Note 1 Hz = 1 cycle per second, 1 kHz = 1000 cycles per second, 1 MHz = 1,000,000 cycles per second, 1 GHz = 1,000,000,000 cycles per second.

The birds 3

If you decide to invest in a TVRO terminal, a whole new range of entertainment is immediately available. Almost anywhere in the world you set up the terminal there are many programmes for you to watch. However, as this book is aimed mainly at Great Britain we will concentrate on the satellite programmes which can be received here.

With a basic system you will normally be able to watch only one satellite at a time, but later in the book we will explain how the system can be adapted to enable a choice of satellites and a choice of the transponders on those satellites.

If you are interested mainly in having more than the usual four domestic TV channels then your first choice will probably be to look at *Intelsat VA F11*. This particular bird is in a position in the Clarke belt at 27.5 degrees West. The programmes carried are all in English. *Children's channel* operates every day from 7 o'clock in the morning until 3 o'clock in the afternoon. The programmes are designed to be watched by very young children and their parents, and range from cartoons, stories to activity games. Older children have serials, cartoons, films, etc. After 3 o'clock the channel becomes *Prem1ere*. This runs through until the early hours of the morning and shows recent films. There are six films every day, which are repeated at different times over a period of several months. There are over a dozen new films introduced every month.

There is another film channel on this satellite which is called *Mirrorvision*. This operates from early afternoon until the early hours of the morning and shows more films. The films are recent productions and, like Prem1ere, they too are repeated at intervals over a few months, so you are able to pick and choose what and when you view.

A programme called *Lifestyle* is on another transponder on this satellite. This is a daytime programme, aimed mainly at the ladies, and consists of 'Agony Aunt'-type programmes, shows about babies and children, health programmes and pet animal shows. It also contains some selected programmes from Yorkshire Television, this company together with W H Smith & Son being the programme provider.

Evening hours on this transponder are filled by a programme called *Screensport,* which, as its name suggests, is for sport enthusiasts. The six hours each evening consist of all the sports currently on terrestrial TV channels plus many others not normally seen in the UK, such as women's wrestling, polo, squash, baseball, and many others.

The only other programme currently on this satellite is a twenty-four-hour news service direct from America, called Cable News Network (CNN). This is a world-wide news service with contributions by satellite from almost anywhere in the world. The programme is presented from studios in Atlanta, Georgia, USA, from where it is uplinked directly to Intelsat VA F11, and then down to the footprint which covers Northern Europe. With this programme you see the news as it happens, not the usual recording of events which is included in the normal news time-slots of the BBC or ITN.

There are two other English language programmes planned for this satellite, and these are a Leisure Channel and an Arts Channel.

From the above information you will see that it is possible to extend the choice of your viewing channels from four to ten, all English speaking.

You may decide that programmes in other languages might be an interesting change, particularly if you are interested in learning another language. The most popular satellite to give you the most options in this direction is *Eutelsat–1 F1,* which is positioned at 13 degrees East.

The Italian domestic equivalent of BBC1, RAI Uno, is transmitted on this bird. This programme consists of News, Classical Music programmes, Religious and Historical programmes, a nightly general interest magazine programme and many live 'spectaculars' which may go on for a couple of hours on a Saturday night. There are shows with dancing girls, jugglers, singers and musicians, reminiscent of the colourful theatre shows like Moulin Rouge and the Folies Bergère. Religious programmes also feature prominently on RAI Uno.

Europa is another channel on Eutelsat–1. This is a mixture of programmes from the Eurovision network. It also includes programmes supplied by EBU member countries. The main language is English, but some of the programmes are in Dutch.

TV5 is a French language programme which operates each evening from 6 pm for about three and a half hours. The programme is a mixture of items from the three national networks of France plus contributions from Switzerland and Belgium. The evening closes with a recording of the main evening news from TDF–1, the French equivalent of the 'Nine O'Clock News'. One point to note regarding TV5 is the fact that the colour standard used for this transmission is the normal system used in France, Russia and a few more Eastern Bloc countries. It is called SECAM and is technically different from the European system of PAL or the American NTSC system. If you watch these programmes on your normal TV the pictures will appear as black and white. It is becoming easier nowdays to purchase multi-standard receivers and if you wish to see the French and Russian pictures in colour you will need such a set. Whilst on the subject of colour standards, it is worth pointing out that the various American programmes referred to so far, CNN on Intelsat VA F11 and the programmes on Eutelsat–1 F1, are in fact converted to the European PAL standard before being downlinked so these programmes are seen in colour on your normal TV.

Earlier in the day, on the same transponder as used by TV5, there is a programme called *Worldnet.* This is a product of the United States Information Agency and consists of live news from Washington DC, scientific programmes and reviews of the previous day's news as seen by ABC, NBC, and the main networks. This programme, unlike TV5, is transmitted in the PAL standard so is seen in colour on European receivers.

There are two other television programmes on this polarisation on Eutelsat–1. (Polarisation will be discussed later in the chapter dealing with dishes.) These programmes are only mentioned in passing because they are encrypted so that only authorised people can receive them. One is called *New World Channel,* which is a multilingual religious programme service, and the other originates in the UK and is called *Sky channel.* In order to resolve these programmes for use it is necessary to have suitable decoding equipment and this is only obtainable by arrangement with the programme provider. Such decoders are generally not made available for use by the home TVRO user, although it is possible that cheaper decoders will soon be manufactured, and it is rumoured that Sky channel may eventually make these available to some users, other than the large cable companies who are the present clients.

There are still a number of other programmes available on Euetlsat–1, such as *Teleclub.* This originates in Switzerland and consists mainly of films, nearly all of which are dubbed into German, or originate in German. This is a pay service designed for German-speaking cable networks and it operates for several hours each evening.

Next we have a film channel originating in Holland called *ATN/Filmnet.* With this service the films are usually in English, with Dutch sub-titles, although there are a few Dutch films without sub-titles. As with Teleclub this service operates for several hours each evening, tending to continue into the early hours on some evenings, particularly at week-ends.

During weekdays the Filmnet Channel is used by *World Public News* (WPN), which is a similar programme to Worldnet on the opposite polarisation. Everyday at 1 o'clock GMT there is a half-hour live transmission of news from Washington DC. Earlier at noon there is a recording of the previous day's news transmission, and then there are numerous programmes, mostly in English, from various countries like Sweden, Jordan, Holland, etc. There are teleconferences from time to time, linked in Belgium but with contributions from many countries including the USA, Holland, Sweden and even the UK.

Further up the band is a programme service from Germany, in German. This is *Sat 1*. This is a mix of programmes, including nightly news bulletins, weather reports and forecasts. There are audience participation shows like Countdown, magazine programmes, pop music shows and there is a nightly quiz where a viewer is telephoned to answer a question. A correct answer wins the viewer 1000 dM, and if it is not won the amount increases by 1000 dM each night until a correct answer is received.

The final programme on this satellite is called *Music Box* which originates in England and consists of 18 hours a day of the latest in pop videos. This is a very popular channel with teenagers and devotees of the pop scene.

The preceding list deals with the most popular satellite programmes which are currently available.

More and more satellites are planned for the future, one of the next to be launched Eutelsat–1 F3 is scheduled to have channels for Spain, Denmark, Sweden, Norway and Turkey, plus another two channels for the UK.

One other satellite which is popular with a lot of people in Europe and the UK is a Russian bird called *Gorizont 7*. This carries the *Moscow 1* programme, the equivalent of BBC1. It also carries the Intersputnik and OIRT TV unilateral programme exchanges, and it carries excerpts from other USSR television programmes from time to time. There are some snags with this particular satellite. The main one as far as we are concerned is the fact that it uses a lower frequency than all the ones mentioned so far. In fact, it uses frequencies which lie in the band which is used by the majority of the USA cable service systems. The other snag is the colour standard is SECAM and not PAL, and in order to conserve the power in the satellite the sound component of the programme is compressed. In order to receive good sound and pictures from Gorizont, therefore, it is necessary to have a different receiver, one which takes account of the compressed sound and expands it again. And you need a multi-standard or SECAM monitor to resolve the colour. Having said all this, if you watch Moscow 1, just as the Russian people do, you can enjoy a lot of cultural programmes, music, ballet, art, etc. There are quite a lot of historical programmes,

Figure 3.1 A selection of satellite TV logos.

poetry, chess and ice-hockey plus a large slice of circus and ice-skating. The programmes are not designed for Western eyes and ears. The language is different and captions look odd to Western eyes, but it does give an insight into the Russian way of everyday life.

The quality of all the programmes mentioned so far is excellent. If the terminal is correctly installed, new viewers are always amazed when they first encounter satellite TV.

If you get interested and want to play about with your installation it is surprising what you can receive. Even if you are outside the normal footprint is is often possible to receive programmes from many places in the world. I have seen programmes from Argentina, Saudi Arabia, Spain, India and even the American Forces Radio and Television Service received on terminals in England and Wales.

There are more than a dozen satellites which are accessible from the UK which are carrying television programmes of one sort or another, and the numbers are growing all the time.

Although most of the information contained in this book is aimed at the viewers in Europe, and the UK in particular, it is probably worth mentioning at this stage that Satellite TV is available in all parts of the world.

The location of the receiving site will determine which satellites are 'visible' and therefore which programmes can be received. Some locations will require tuners and equipment for the C-Band signals, which are used for most of the satellites serving the USA and the USSR; others will need Ku-Band facilities to receive the systems used in Europe. Just as here in the UK, it is possible to receive some C-Band and some Ku-Band transmissions, so it is in many other parts of the world.

Here is a short list of some of the Satellite TV programmes which can be received in widely separated regions. The size of dish required for satisfactory reception will vary from place to place.

Australasia
CNN from Los Angeles
Australian Broadcasting Co. from Brisbane
Australian Broadcasting Co. from Perth
JISO from Japan
AFRTS from America
RPN 9 TV from Manila
TBS 5 TV from Bangkok
Warner 2 from Kuala Lumpur

South Africa
Marocaine TV
Argentina Televisora Color from Buenos Aires

Embratel from Brazil
Saudi Arabian TV
TV Nigeria
Sudan Television
TV Manila
TV Bangkok
TV Malaysia
Algeria TV
Oman TV
TV Chile
Spain TVE1
Spain TVE2
AFRTS from America
TV Portugal

Indonesia
TV Brazil
Saudi Arabian TV
Niger TV
Sudan TV
Zaire TV
Australian Broadcasting Co.
AFRTS from America

USA
Some details about the vast range of American programmes available on satellite are given later in the book. In addition to the 'home produce', it is also possible to receive a number of foreign programmes, including:
Mexico TV
TV Chile
Venezuela
Argentine
Brazil
USSR
Saudi Arabia
Sudan
Zaire
TV Portugal
RAI Italy

4 The terminal

Let us now discuss the equipment which is necessary for a TVRO (TeleVision Receive Only) terminal, starting with the simplest and consequently cheapest installation.

The first item to consider is *the dish*. Because the frequencies used for the majority of satellite services which are receivable in Europe are considerably higher than the ones used by the Americans, the size of the dish is smaller in Europe. Whereas it is common to see TVROs in the United States using 10 foot, 15 foot (and even larger) diameter dishes, the size of dish required in the UK for an excellent performance averages about 6 foot (1.8 metre). It is possible to receive satisfactory pictures, in certain locations, with a dish of diameter 1.2 metre, but one of this size can be affected by bad weather conditions, such as snow. The recommended diameter is 1.8 metre.

The dish acts as a collector for the very weak signal coming back to Earth from the satellite. The efficiency of the dish is dependent on the accuracy with which it is made, as its shape has to be a true parabolic curve. A solid dish, usually spun out of aluminium, or similar material, is considered to be the most efficient. Careful manufacturing methods are, however, now able to produce sectional dishes which are efficient enough for our purpose. The sectional dishes are made either in two halves, which can be bolted together on assembly, or they may be made up of six panels, which are also bolted together on site. For obvious reasons it is easier to transport a sectional dish rather than a solid one. There is another type of dish which is now emerging on the scene, and this is one made up of wire mesh. The Americans have had mesh dishes for many years. But because of the different frequencies, the European equivalent has to be made of much finer mesh. One advantage of the mesh dish seems to be aesthetic. If it is sited in a garden, for instance, it is less obvious because the garden shrubbery can be seen through the dish.

Having decided on a dish, we then have the problem of supporting it. There are many varieties of mount available, varying from tripods and frames to stand in your garden, to wall brackets and frames to mount the assembly on a flat roof. Your location may determine which type is the most suitable.

This is probably a good point to say a few words about the site requirements. First you must be able to 'see' the birds. As they are all in position above the equator the first requirement is that you should have an unobstructed view to the South if your site is in the northern hemisphere, and an unobstructed North view if you are in the southern hemisphere. Secondly the angle above the horizontal that you can see needs to be at least 10 degrees or so. The elevation of the

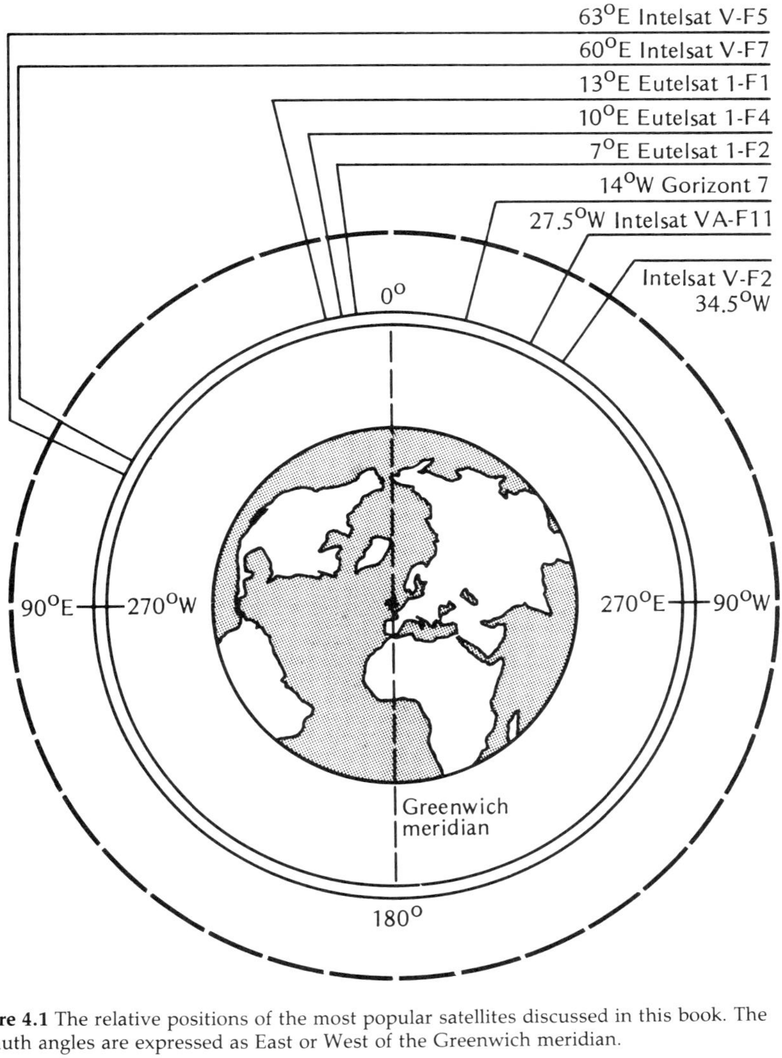

Figure 4.1 The relative positions of the most popular satellites discussed in this book. The azimuth angles are expressed as East or West of the Greenwich meridian.

two most popular birds in the UK are in the region of 28 to 30 degrees elevation, dependent upon the exact location of the receiving site (Figure 4.1).

The wind loading of dishes needs to be considered when choosing your equipment. The larger the size of the dish, the more wind resistance it has. It is always wiser to position the assembly at ground level wherever possible. It is also desirable to fix the dish mounting base with bolts to a prepared concrete base, or to weight it down with gravel or heavy stones.

In order to line-up the dish with the satellite, it is necessary to be able to move it in bearing, and in elevation. The simplest form of mount is known as an *az/el* mount for this reason. This mount consists of a frame onto which is bolted the dish. Information is available which gives the azimuth bearing of each satellite, as viewed from your location, so with the aid of a simple compass it is possible to point the dish in the correct general direction. A device known as an inclinometer is then used to adjust the elevation angle of the dish to the required satellite elevation. It follows from this that if you need to adjust the dish for the reception of other satellites you will have to move it in two planes each time, i.e. azimuth and elevation.

Just as with optical telescopes, the dish mount designers have now come up with a different type of mount, called a *polar mount*. This mount consists of a frame and vertical pole, but the dish is attached with its axis in line with the axis of the Earth. Without going into too much technical detail a polar mount, when correctly installed, permits you to scan the whole of the Clarke Belt with only movement in one direction, rather than the two required by the az/el mount. The polar mount, therefore, has a distinct advantage if you wish to change satellites from time to time. A further advantage is that it is possible to have a motor-drive to change this position. To motor-drive an az/el mount would require two motors and much more complicated control systems to operate them.

We are still dealing with the first requirement, the dish, and already it is clear that there are numerous variations available, all of which allow you to expand the original basic system, as interest develops.

Having obtained the dish we now need to start on the electronics. The signals collected by the dish are brought to focus, just like the lens of a telescope, at a point in front of the dish, known as the *prime-focus point*. This is the position where we must fix a device known as a *feed-horn,* a small metal device which is usually bolted on to a box which contains the first lot of electronic circuitry. This box of electronics is called either an LNA, an LNB or an LNC. The most popular device is an LNB, which is a Low Noise Amplifier and Block Downconverter. This device amplifies the very weak signal received by the dish and then changes its frequencies to a lower section of the frequency spectrum. This enables the signals to be fed along a piece of coaxial cable, not too dissimilar to the aerial downlead on your television set, and into the next section of the TVRO terminal, the tuner demodulator. This tuner is normally inside the house, perhaps sitting on top of your TV set. The power needed by the LNB, out at the dish, is normally supplied by the tuner unit and fed to the LNB on the same coaxial cable which carries the radio frequency signals from the LNB to the tuner.

To recapitulate on the requirements for a simple TVRO, you require a dish pointing at one satellite. You need an LNB mounted on the dish, and you need a tuner in the house with a coaxial cable connected to the LNB (Figure 4.2).

Initially a simple installation is an exciting new 'toy', but most people are not content to just receive the few programmes which are available on one polarisation from one satellite, and they are soon to want more flexibility in the system. A satellite TV terminal is rather like the construction sets which were more popular years ago, like Meccano, where you first bought a size number 1, and then became more ambitious and added a 1× to make it into a size 2, then a 2× and so on. The TVRO can be dealt with in a similar fashion. Perhaps the first addition should be a second LNB, set to the opposite polarisation to the original. This would give you immediate access to the rest of the transponders on the satellite you are looking at, so you might double the number of programmes at a stroke. In order to select alternative LNBs you simply change the cable into the tuner. A more sophisticated method is to have a relay operated switching device so that you can change polarisation from a switch on the tuner, and an even more sophisticated arrangement which is now becoming available is a device called a polar-rotor. This is generally a flexible membrane which is fitted to the single LNB that is moved electromagnetically into the two positions required. Again this is operated by a switch on the tuner. There is yet another alternative, and that is to mount the LNB with a small motor drive to actually rotate it through the 90 degrees from one polarity to the other. All the above methods are equally effective and your choice might be dependent on cost or the method you prefer.

Having equipped yourself with the ability to view all the transponders on the satellite, whatever their polarisation, you will now start to think about looking at other satellites. This means, generally, the ability to move the dish. In the case of the cable companies and professional bodies who might want to look at more than one satellite simultaneously, it means the installation of more dishes, each one directed at a different bird. A home TVRO normally only needs to see one bird at a time, so it is usual to move the dish. The cheapest way of achieving this is to paint marks on the dish

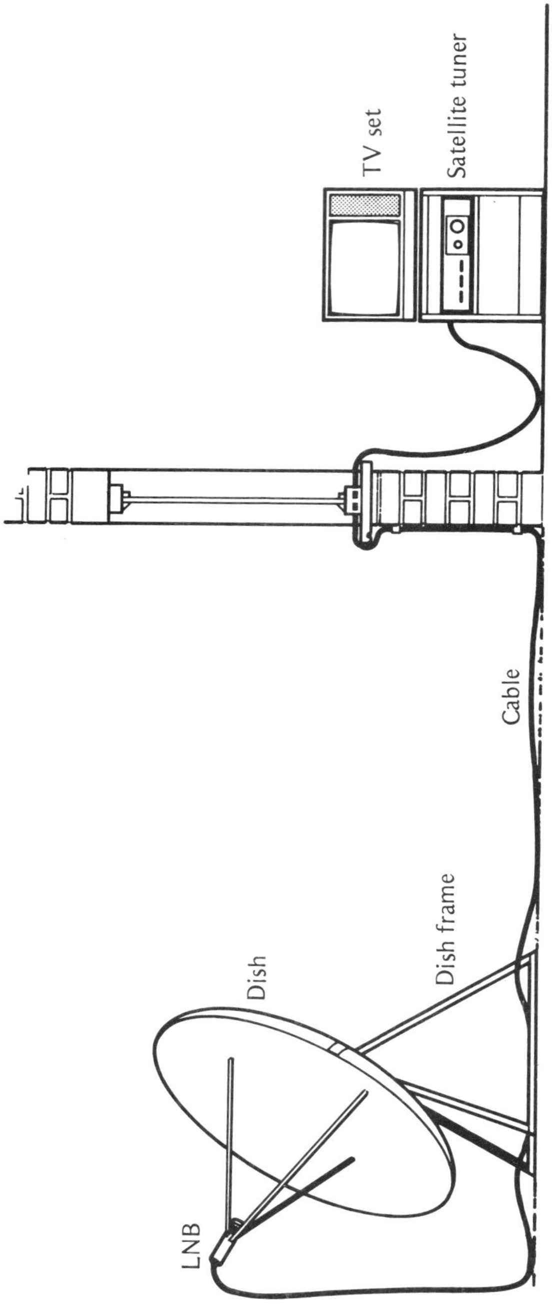

Figure 4.2 The basic installation. In its simplest form it is only necessary to have one cable from the dish to the satellite receiver in the house.

mount, perhaps with coincident lines, to show the bearing and elevation of the various birds. The simplest arrangement is to have the dish motor-driven. If you think you might want eventually to have the dish motorised, it is worth-while to consider a polar mount dish right from the start. A polar mount is rather more time consuming in its initial setting-up phase, because the alignment needed to ensure accurate tracking has to be very precise. Having decided on a polar mount, however, it is possible to move from bird to bird with one simple movement, and the addition of one motor, with a position indicator control box inside the house makes it very easy to change. The next development, which is already beyond the drawing-board stage, is a computer-type programmable chip, which will allow the satellite you want to see to be located automatically, having been selected on your tuner.

To sum up this Chapter, a TVRO terminal consists of a dish, fitted with suitable electronics in the shape of LNBs, a length of high-grade cable to connect into a tuner unit which can then be plugged into your normal TV set. An alternative to the tuner and normal TV is the use of a TV which has suitable circuits built into it to enable satellite reception in addition to terrestrial television. It is a matter of personal choice, but I favour the separate tuner and normal TV as it seems to allow more flexibility, such as using a videotape recorder, which is not easy with the special TV, and of course you are not limited to the one set if you use a separate tuner for satellite.

Having just mentioned a videotape recorder it is worth pointing out that, with the separate satellite tuner installation, it is a fairly simple matter to tune one of the buttons on the VCR to the satellite signal so that it is possible to record programmes from satellite (Figure 4.3). Another way to achieve this is to use the video and audio outputs from the tuner unit to feed the satellite programmes permanently into the VCR, and then simply use the r.f. modulator in the VCR to send the satellite programmes into your normal TV in the same way that you would select recording tapes.

Finally a word or two about the various types of tuners which are currently available. Tuners are now quite numerous, some are more sophisticated than others, and performances do vary. The majority of tuners available are not manufactured in this country. I am strongly of the opinion that British is best, and I would recommend that you compare what is available before finally deciding to buy. The method of selecting the channels varies from tuner to tuner. Some are like a radio set where you simply turn a knob to move up or down the band. Others, however, have channel indicators, normally illuminated numbers, which can be programmed to the various channels available and then the operation of a switch will change the channels. Even a hand-held remote control system is available with one of the tuners.

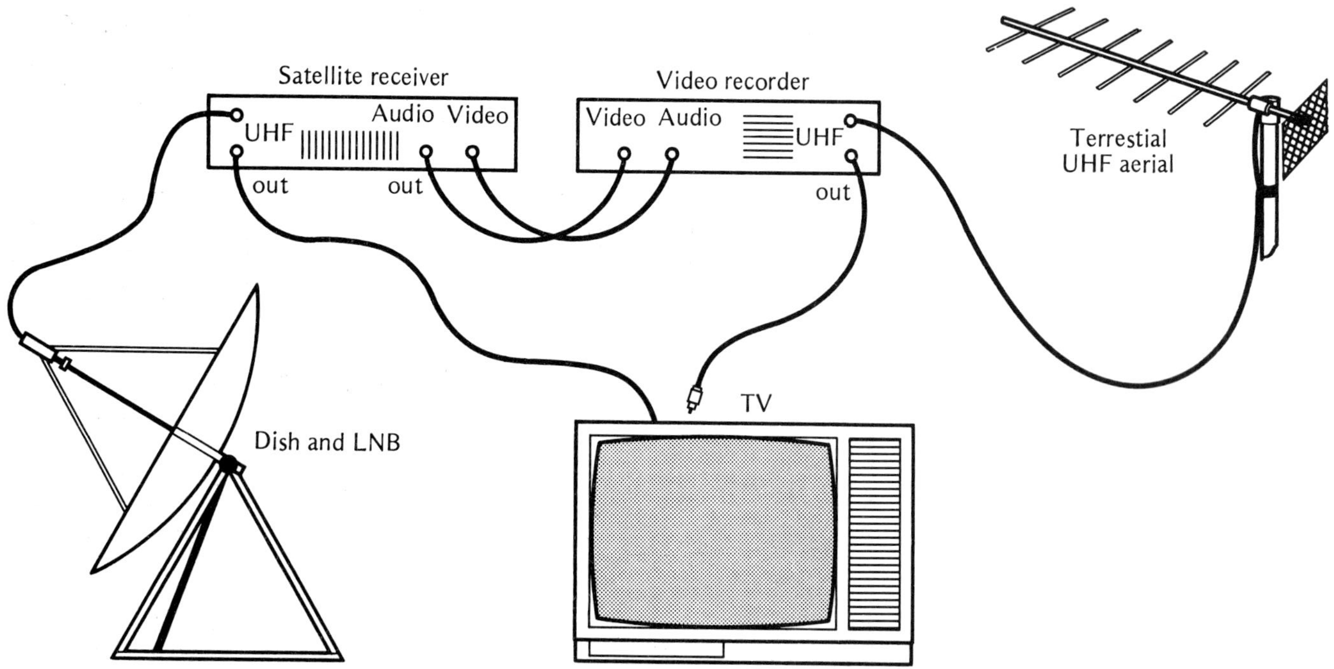

Figure 4.3 Typical equipment layout. To save constant unplugging at the TV aerial input, an alternative arrangement is to leave the VCR permanently connected to the TV set and select your viewing and recording channels by using the tuner on the VCR and/or the tuner/line switch on the VCR.

Figure 4.4(a) SATVRN domestic TDM1200 series satellite tuner with remote control.

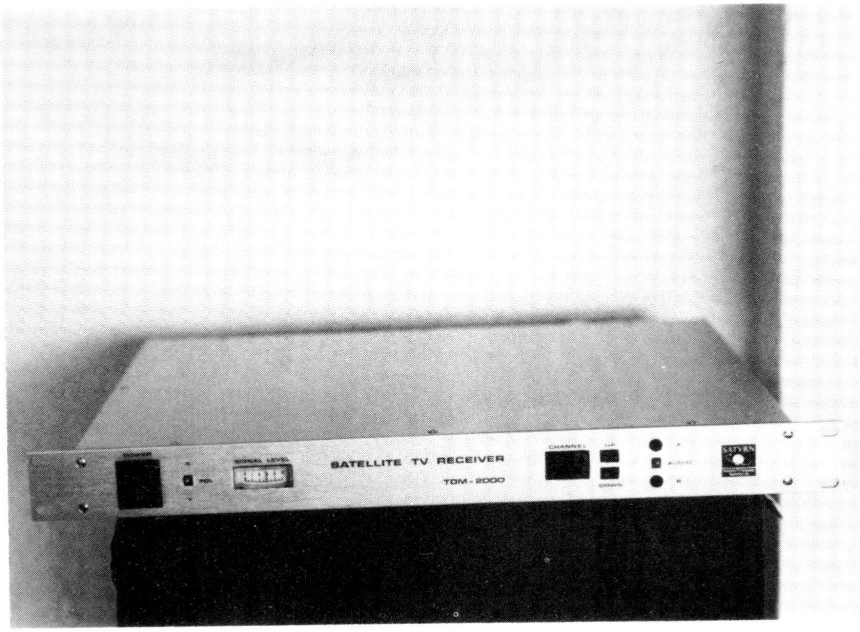

Figure 4.4 SATVRN professional TDM2000 series satellite tuner for rack mounting.

There is one other point to consider when choosing your tuner. The novelty of satellite television often masks the fact that the sound quality on some of the services sounds different to normal terrestrial TV sound, and it soon begins to grate. Some of the tuners at present on the market do not make provision for the different sound standards, so it is worth asking about this factor before deciding to buy.

There is now another method of obtaining a TVRO system, and that is to rent one. A few of the TV rental companies are now renting satellite terminals. DER is just such a company. At the present time (1986) the initial payment, which includes installation, receiving licence and the first year's rental plus the first year's subscription to the programme providers, is about £750. The exact amount of annual rental, thereafter, is in the region of £180 at present.

Thorn-EMI, under the name of Ferguson, have also entered the TVRO market, and they are selling complete terminals through a number of their dealer outlets. The dealers can contact expert installation teams, if required, and the whole thing comes as a package deal. The prices for this terminal is in the region of £1250 at present, but like all the prices in this particular field they are expected to drop.

Costings 5

One of the important factors, if not *the* most important, when considering the installation of a TVRO system is how much it is going to cost.

As the price of equipment is constantly changing it is difficult to be specific. However, I will try to give some guidance from the figures I have available at the time of writing.

The simplest system commercially available is a 1.5 metre one-piece spun aluminium dish, suitable for standing in the garden or on a patio, a low-noise LNB for mounting on the dish, a length of coaxial cable, approximately 20 metres, and the tuner unit for standing on top of the TV set, complete with infra-red hand held remote control, plus a licence from the Home Office, will cost you £995.

If you think you that might want to expand the system, it is worth considering an initial installation which uses a 1.8 metre dish. This is a more adaptable system for future development but it will put the initial cost up to around £1250.

Having used the installation for some time, you will probably want to expand the system so that you can receive all the programmes from a particular satellite without having to go out into the garden to turn the LNB through 90 degrees. This can be done by adding a second LNB to the dish. This, together with the necessary waveguide, will cost you around £850.

Supposing you want to be even more sophisticated and be able to look at other satellites, then you must be able to repoint the dish. This is not easy if you have the simple patio-type mounting. What you require for ease of movement is a polar mount for the dish. Not only is it easy to move from bird to bird with this type of mounting, it is also a comparatively simple matter to fit a motor drive to the dish and operate remotely from inside the house. At the time of writing, a polar mount for the 1.8 metre dish will add about £600 to the initial installation, and the further addition of a motor drive would cost something in the region of £350 to £400.

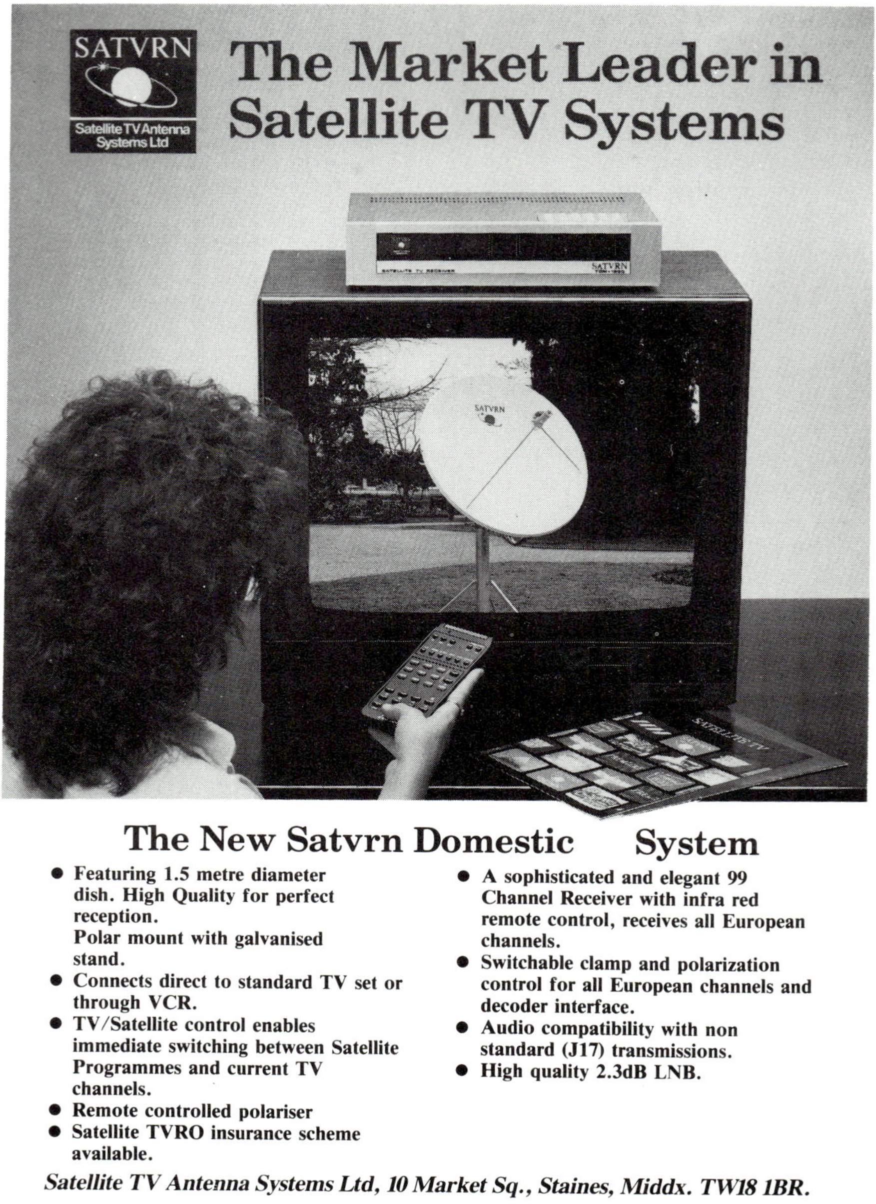

Figure 5.1 Copy of SATVRN advert.

To make the system really armchair controlled, you might also like to add the necessary switching equipment to enable you to change the LNB polarity. This can be done in two ways. One is to switch between two LNBs on the dish, and the other is to use one LNB fitted with a device called a polar-rotor. This is a flexible membrane fitted to the single LNB which can be changed to the opposite polarisation by means of an electromagnet. As far as the turner unit is concerned it does not matter which system you choose. The tuner unit has a selector switch to enable polarity changes. The cost of having a relay box to change between two LNBs is in the £200 region. This may sound expensive but, although I called it a relay box, it is in fact a very complicated piece of electronic circuitry, because it is not feasible to use a simple relay for this purpose, owing to the very high frequencies which are involved. At the present time the cost of the polar-rotor system is cheaper than the two LNB arrangement. However the polar-rotor is very new on the market and it will almost certainly prove to be the most popular option and so it is likely to become significantly cheaper than the two LNB system.

Having gone through the various options, there may be some of you who would like to have the ultimate system right from the start, so let us consider what this would cost. A 1.8 metre dish, installed on a polar mount, complete with motor drive, a polar-rotor LNB, just one coaxial cable from the dish to the tuner, all complete with a hand-held remote control for the tuner and a control box for seelecting which satellite the dish is aimed at, and we are talking of equipment costs, at today's prices, of something in the region of £3000. If you are going to have a system like this, then you will probably want it professionally installed. The installation costs are dependent on many factors, such as where the dish can be positioned, how long a cable is required, etc., and each location really needs a site survey in advance to ascertain these points. As a guide, however, the average installation costs vary between £200 and £500. It is worth while to shop around the various dealers offering this service, and then decide which suits you best.

There are other costs to consider, in addition to the equipment and installation, you require a licence from the Home Office to permit the reception of satellite transmissions (Figure 5.2). This is easily obtained from the Home Office, and it is a once-only payment of £10.

The other items which need paying for are some of the programme channels. These are the film channels like Prem1ere, Mirrorvision and Telecub. The method of collecting these payments is still being considered in most cases. However, there is a company called Galaxy Television, who are appointing dealers to collect their fees. Prem1ere and Music Box are part of the Galaxy package, and prices for these services is in the £8 to £9 per month range at present. Prem1ere on its own costs £7.50 per month. Mirrorvision is currently £7.95 and the

Department of Trade and Industry
Radio Regulatory Division

Application for a Licence for Television Receive Only Satellite Receiving Equipment (TVRO)

- Please read the notes before completing the form overleaf
- Please complete the form in block capitals in dark ink or typescript

Notes

1 Under the Wireless Telegraphy Act 1949, a licence is required to install any satellite receiving equipment. This form should be used to apply only for a licence for Television Receive Only satellite receiving equipment (TVRO).

2 The licensee will require a separate licence for each TVRO receiving equipment location.

3 The licence is confined to the reception of television or sound radio programmes forming part of a programme service intended ultimately for reception by members of the public or restricted groups or audiences.

4 The issue of a licence does not mean that reception will be feasible. Transmissions of television programmes from low powered telecommunications satellites are not protected from interference from other authorised radio services, existing or future. It may therefore be impossible to obtain satisfactory reception at a particular site.

5 It is a condition of TVRO licences that the apparatus shall not cause interference to other users of the radio spectrum.

6 The issue of a TVRO licence does not replace or remove any requirements for licences under the Telecommunications Act 1984 or the Cable and Broadcasting Act 1984. In general, satellite television cable systems serving more than one dwelling house or a single set of premises require licences under those Acts as well as a TVRO receiving equipment licence. Guidance may be obtained from the Cable Authority, Gillingham House, Gillingham Street, London SW1V 1HU.

7 The issue of a TVRO licence does not replace or remove the need to obtain all other relevant consents and authorisations. In particular it does not affect copyright in the programmes.

8 All necessary planning permissions should be obtained before the TVRO is installed.

9 The licence fee is £10 and the licence is valid indefinitely. The fee should be submitted with this application to the address shown at the end of this form. Cheques/postal orders should be made payable to Department of Trade and Industry and crossed "A/c Payee".

BR 39 M953

Please give the following details:

1 Full title of the Company, firm or individual in whose name the Licence is required. __________

2 Address (Registered Office if a Limited Company). __________

3 Address for correspondence (if different from above). __________

4 Name and telephone number of person to whom all enquiries should be addressed. __________

5 Address of TVRO receiving equipment. __________

Application is made on behalf of __________
for a licence for TVRO satellite receiving equipment at the address (at 5) above. Any change in any of the details given will be notified to the Secretary of State.

Signature __________ Date __________

Checklist

☐ Have you read the Notes?

☐ Have you enclosed your cheque/postal order?

Please return the form to:

The Cashier
Department of Trade and Industry
Accounts Branch
24–26 Newport Road
Cardiff CF2 1SY

Figure 5.2 Copy of licence application.

Children's Channel is £0.30–£0.60 per month. Not all the satellite programmes available require that you pay a subscription to watch. The free ones include RAI-Uno, from Italy. This programme is advertiser supported, as is Europa and SAT 1. TV5 is a free service, and several of the other services have not yet published any charges. It is an offence to watch programmes which have a subscription charge, without paying, but at present it depends on the viewer's conscience whether they pay or not. If the viewing numbers increase, as I am sure they will, then it will be necessary for subscriptions to be collected in order to ensure a continuance of the programmes.

Various methods for ensuring payment are under consideration, and probably the most discussed is that programme providers should scramble their transmissions. There are many people who think this is not the direction to take, and these people favour advertiser-supported services. Based on the UK Commercial Television system of advertising, one group feel that the large area covered by a satellite footprint will enable vast coverage of an advertiser's product. Under such a scheme the advertiser wants to be seen by as many potential customers as possible, so scrambling is not what he wants. Other factors to consider when scrambling, is the cost of the extra equipment needed, not only by the programme provider but by every receiving terminal. This equipment is not cheap, it adds another possible source of fault liability, and, unless it is very sophisticated, there is a likelihood of some electronic whizz-kids coming along and producing cheaper decoders to beat the scrambler. So, as you see, it is not a clear-cut method and it needs a lot of careful consideration.

If, eventually, some form of scrambling takes place, I think you can rest assured that the home TVRO owner will not be overlooked, and suitable descramblers will be made available to him. In the case of Sky channel, which has been scrambled from the start, there is some talk in the trade that a simple decoder is being considered, and it may well become available to the SMATV and home TVRO user in the not-too-distant future.

6 The programme providers

So far we have dealt with the cost of an installation, and the price of a licence to allow the reception of satellite TV. This Chapter is devoted to the programme providers: why are they doing it, and how are they being financed.

The original concept of comparatively low-power TV satellites, as opposed to the higher-powered DBS satellite systems, was simply a means of providing cable companies throughout Europe with suitable programmes to extend their facilities. Part of the revenue, collected by the cable companies, would be passed on to the various programme providers to finance their operarion.

With the demise of the DBS programme, and the advent of TVROs and SMATVs the problems associated with obtaining revenue from their viewers has produced many headaches amongst the programme providers. A simple solution to the problem does not seem possible, and various ideas to prevent illegal reception of their programmes are constantly being sought. One idea, which seems to be gaining approval by a number of programmers, is to scramble the signals in such a way that they can only be resolved by people who have been approved and agree to pay the appropriate rental for the service.

One of the complications with scrambling the signals is the fact that there is no agreement between the various bodies involved, for a common system of scrambling. At one time the French channel TV5 was scrambled, as is one of the religious channels. The system used here was different to the system used by Sky channel so you needed different descrambling equipment for each service.

As far as the three English channels on Intelsat are concerned, that is Prem1ere, Mirrorvision and Screen Sport, together with the various other programmes on this 'bird', such as Arts channel, Children's Channel and Lifestyle, I understand that the programmers are cooperating to find a scrambling system which would be compatible with all the aforementioned programmes.

A company, called Galaxy Satellite Television, is currently

arranging methods of revenue collection for several of the channels listed in the last paragraph. In a recent document they say "Most of the channels which this company represents will shortly be scrambled. This is to ensure that those who wish to receive the channels pay the appropriate fee. Those people who are currently stealing our signals will very shortly be prevented from doing so. It will be up to us to sell them the idea of paying a fee; given the current prices of decoders, it is likely they will also be rented to our subscribers, possibly by including it in an overall subscription fee. Our encryption system will be addressable. This means that we will have the facility to switch decoders on and off through digital signals inserted into each channel. Similar technology is already being used on mobile telephone equipment. Although those who do not pay will naturally be cut off, the advantage for viewers is that only one decoder will be needed (ultimately built into the TV sets)." The current cost of the Galaxy package, which allows reception of Prem1ere, Children's channel and Screen Sport, is slightly less than £2.30 a week.

It remains to be seen what the ultimate arrangements will be. There are many points of view about the merits of scrambling, and changes are taking place all the time. For example, TV5 used to be scrambled, but it is now transmitted in the clear. SAT 1 and 3SAT from Germany used to be scrambled but they too have gone clear. Several channels now include a number of 'commercials', so they are advertiser-based for some of their revenue, and it may not be in their best interests to scramble the programmes and thus deny some of their advertisers possible coverage. In the USA, on the other hand, the main satellite movie channel HBO (Home Box Office), after several years in the clear, has recently gone scrambled.

Even if, and when, some of the channels scramble their signals, it is likely that there will still be plenty of clear channels available for the satellite TV viewer to watch.

The problem of collection of their dues from the various users is also something that the programme companies are busy trying to resolve. Even if they do not adopt a system of scrambling, the programme providers need to be paid. It is very unlikely that any of them will receive sufficient revenue from advertising, so they will have to devise other methods of payment. One suggestion, currently being considered, is where the supplier of the satellite receiving equipment would add the first year's rental for the various programme companies to his own costs. This money would then be handed on to the companies, together with details of the owner of the installation, and the address. Future payments would then be requested direct from the programme providers to the end user. This system is obviously not foolproof, and it depends, to a certain extent, on the agreement of the equipment installer and the integrity of the user, it does at least provide the programme companies with some information about the eventual consumers of their product.

Some of the programme providers have no provision at all for TVRO and SMATV owners to contribute money for using their programmes. Two of these, in particular, Teleclub and ATN/Filmnet, simply state that their signals are only allowed to be used by authorised cable companies, and they say quite simply that it is illegal to receive their programmes. If these transmissions remain unscrambled it is difficult to imagine that viewers will ignore them and refuse to watch, simply because it is illegal to do so. It would probably be far better for all these companies to recognise that their signals are receivable on simple terminals, and they should be looking for means to receive adequate payment for their service because the number of terminals is growing rapidly. At the present time it is impossible to detect which satellite programme people are watching, unlike the detection equipment used by the authorities, for terrestrial television transmissions, when checking for licence dodgers. If there is something worth watching on these channels then someone will surely be doing so.

The CNN channel is aimed at news organisations and SMATV operators. Individual TVRO installations who might wish to watch CNN are invited to get in touch with David Garland, CNN, 66/67 Newman Street, London, W1P 3LA.

WPN from Belgium is also designed mainly for cable networks. Individuals who wish to use the service should contact Marc de Cock at ITS, Ezelstraat 1, 8000 Brugge, Belgium.

If we take into account the fact that the world is constantly changing, and bearing in mind that scrambling is likely to take place on some of the existing channels, what then is generally available now on satellite TV here in Europe.

Premlere English speaking movie channel.
Mirrorvision English speaking movie channel.
Children's channel Daytime programme for children. Clear at present but will probably scramble soon.
Lifestyle A programme aimed at women viewers.
Screen Sport Sports programmes from around the world in English. Clear at present but will probably scramble soon.
The Arts channel An early morning cultural programme, aimed at cable companies who might wish to record the programmes for later transmission.
Cable news network A round-the-clock American news programme.
RAI–Uno The Italian public number 1 programme, similar to BBC1 with news, cultural programmes, popular and classical music programmes and light entertainment spectaculars. The language is

Italian and there is a fair proportion of religious programmes.
3–Sat A German language entertainment programme which features contributions from German ZDF, Austrian OSR and Swiss SRG TV networks. This signal is not as strong in Britain as some of the other services and a much larger dish, say 3.5 metres, is probably necessary to receive satisfactory pictures.
Europa An entertainment programme, mainly in English, but sometimes Dutch and other European languages, transmitting programmes provided by member countries of the EBU (European Broadcasting Union).
TV–5 A French language programme, although the evening's menu is introduced bi-lingually in English and French, with programmes contributed by the three main French networks, in addition to programmes from the Swiss and Belgian network services. This programme is transmitted in the colour standard adopted by France and the USSR, and is only seen in black and white on the normal European TV set. In order to see the pictures in colour, it is necessary to use either a multi-standard receiver, or a SECAM receiver or monitor.
New World channel This is a religious programme channel which is scrambled. It originates in Norway and presents multilingual programmes.
Worldnet This channel is part of the US Information Agency, and it transmits news, documentaries and educational programmes from the USA.
Sky channel An English language entertainment network, of sports, movies, children's programmes and documentaries. This is a scrambled programme and decoders are only issued to approved cable companies at present, although decoders may eventually become available for other authorised users.
Teleclub Premium movie channel, mostly German language, or films dubbed into German, which originates in Switzerland.
RTL+ Luxembourg German-language public TV channel 7, featuring news, entertainment and films. This is an advertiser-supported channel, but like 3–SAT it requires a large dish for good reception in Britain.
ATN/Filmnet Originating in Belgium this is a premium film channel, mostly English speaking films with Dutch sub-titles, but with the occasional Dutch film shown without sub-titles.
WPN World Public News originates in Belgium. It transmits live news from Washington DC every weekday, together with programme contributions from Jordan, Norway, Finland, Holland, etc. Most of the programmes are in English.
Sat 1 A German cable service which has news, documentaries, magazine programmes and films. The main language is German but

occasionally American or English music programmes are shown in the original language.

Music Box Pop music videos make up the bulk of this English-speaking programme which runs for about 18 hours every day.

Moscow 1 Like RAI–Uno, this is the normal domestic TV programme from Russia. Also like TV5, the colour system is SECAM. It is also necessary to have special circuitry in the receiver to reproduce the correct sound quality. A lot of classical music programmes, ballet and circus are transmitted on Moscow 1, also a fair selection of sport, ice-skating and documentary programmes, together with extensive news and current affairs programmes. The Russian language is not easy to follow, but the gist of the programmes is often understood.

More choice at a price 7

We have so far dealt only with the most popular, and economical, installations, where the dish size has been 1.8 metres or less in diameter. If you are able to obtain and install larger dishes, however, it is possible to extend your choice of satellite programmes considerably.

Cost is a rapidly expanding factor, where dish size is concerned. A 7 metre dish will probably cost in the region of £25,000, and also will require planning permission to install.

Here is a list of some of the currently available satellite TV programmes, together with the estimated size of dish which is required in the UK to receive them.

Moscow 1	1.8 metre
RAI Uno	1.8 metre
3-SAT	3.5 metre
RTL+	3.5 metre
France TF1	3.5 metre
Algeria TV	4.5 metre
Saudi Arabia TV	4.5 metre
ATC Argentina	4.5 metre
Libya TV	4.5 metre
Nigeria TV	6.0 metre
American Forces RTS	6.0 metre
Morocco TV	6.0 metre
Spain TVE–1	7.0 metre
Spain TVE–2	7.0 metre
Oman TV	7.5 metre
Niger TV–Sahel	7.5 metre
Sudan TV	7.5 metre
Zaire TV	7.5 metre
Chile TV	9.0 metre

The list above gives some idea of what is available, given a large enough dish. There are some subtle differences in some of these signals however. For instance the AFRTS service from the USA does require that you make slight adjustments to the dish position from time to time because the orbit of Intelsat IV F8, which carries this service, is an inclined geostationary one and it does require a small degree of tracking. This signal may not always be available in the UK because this satellite is below the horizon from time to time. The Saudi-Arabian transmission on Intelsat IVA F4 uses two separate transponders, one for the vision signal and another for the sound, so you require two tuner units to receive the complete signal.

It is probably worth mentioning at this point that most of the services listed above are transmitted in the part of the frequency spectrum which is used for the bulk of the American satellite TV signals. This is in the 4 GHz band. All the programmes referred to earlier in the book, with the exception of Moscow 1, are in fact transmitted in the 11–12 GHz band. Separate tuners and LNBs are required for each system. It is envisaged that any future satellite TV transmissions will be in the higher, 11–12 GHz band. Some of the more recent services for the Americas are already being transmitted in this higher frequency range.

Satellites are not only used for the transmission of complete programmes, they are also used to convey unedited material from one country to another. These are called *unilateral transmissions* and they are used mainly for news stories, which can then be edited and included in the general news buttetins around the world.

There are daily exchanges of news, by satellite, between the member countries of the EBU, as there are between the Warsaw Pact countries of the Eastern Bloc. Similarly there is a regular supply of news items sent from France to Catalonia, and this is sent on the same transponder that is used in the evenings to transmit the TV5 programme. Unlike the TV5 programme, however, this is transmitted in the European colour standard PAL, so it is seen in colour on European receivers.

As these services are not designed for the general public it is probably an offence to disclose anything you might see. However, it is another interesting aspect of satellite TV which is often quite interesting to watch. In the USA, where they operate a 'free sky' policy, there is a growing cult following for many of these unilateral transmissions. There is a certain feeling of excitement if you happen to drop in on one of these unilateral news feeds. It often happens that things go wrong during transmission and several re-takes can be necessary. I recall seeing pictures from Moscow some time ago, when they were setting up for a news item for Germany. The scene showed the German correspondent in front of the large photograph of Red Square. Cameramen and scene-hands kept appearing in shot and the commentator got more and more frustrated as he had to keep repeating his piece. It was reminiscent of some of the out-takes which are featured from time to time in programmes like 'It'll be alright on the Night'.

8 The USA and Canada

In Chapter 1 it was suggested that in America, the home TVRO installation was probably the fastest growth industry at the present time. Cable is also a very popular system of Television distribution, and so are sound programmes. So, having discussed satellite TV as it affects Europe, let us look at the USA and Canada.

First we will consider the cable companies. Table 8.1 shows what a vast choice of programmes is available to subscribers to the Rogers Cable TV system in Ontario.

Companies, such as Rogers, have TVRO terminals at their headquarters. These usually comprise several dishes focussed on different satellites. It is from such terminals that the cable companies obtain the programmes which they then distribute to their subscribers by cable.

Table 8.1 Rogers cable services TV programmes

To receive	*Channel*	*Dial*
2	WGRZ(NBC)	13
3	CKVR(CBC)	15
4	WIVB(CBS)	05/16
5	CBLT(CBC)	06
7	WKBW(ABC)	09/17
9	CFTO(CTV)	08
10	CTEN Community Programme	10
11	CHCH(Ind)	11
12	CHEX(CBC)	18
13	CKCO(CTV)	19
17	WNED(PBS)	20
19	CICA(TVO)	02
22	CKGN(Global)	03

Table 8.1 *continued*

To receive	*Channel*	*Dial*
25	CBLFT(CBC)	12
29	WUTV(Ind)	21
47	CFMT(Ind)	04
57	CITY (Ind)	07
SH/GOVT	Showcase/House of Commons	14
TV	TV Listings	22
A & E	Arts ans Entertainment	23
FC	First Choice/Supercha	24
CNN	Cable News Network	25
TSN	The Sports Network	26
TNN	The Nashville Network	27
TLC	The Learning Channel	28
MM	Much Music	29
TL	Telelatino	30
SE	Super Ecram	31
CV	Chinavision	332
MTV	MTV II	33
CTY	CITY II	34
BN	Broadcast News	35
SM	Stock Market	36
W/A	Weather/Airport	37
FS	Future Service	38
FS	Future Service	39
FS	Future Service	40
FS	Future Service	41

Although this book is mainly dealing with satellite TV it is appropriate at this point to show what is available on the same Rogers Cable Service, as far as radio programmes are concerned. The following list shows what is available.

Table 8.2 Rogers cable services radio programmes

Frequency	*Station*
88.7	Arts and entertainment
89.1	First choice/Superchannel
89.5	Nashville

Table 8.2 *continued*

Frequency	*Station*
89.9	Country music TV
90.3	Much music
91.5	CJRT Toronto
91.9	CKPC Brantford
92.3	WNED Buffalo
92.7	EDUCAT Toronto
93.1	WBUF Buffalo
93.5	CHAY Barrie
93.9	WBLK Buffalo
94.5	CBL Toronto
94.9	CKQT Oshawa
95.3	WDCX Buffalo
95.7	CKDS Hamilton
96.1	WJYE Buffalo
96.9	CKGL Kitchener
97.7	CJQR St Catharines
98.5	CHFI Toronto
99.5	CKO Toronto
100.3	CKFM Toronto
101.1	CHIN Toronto
101.5	CJBC–AM Toronto
101.9	WGRQ Buffalo
102.5	CFNY Toronto
102.9	CFMX Toronto
103.3	WPHD Buffalo
103.7	CHCR–GRK Toronto
104.1	CHIR –GRK Toronto
104.9	CHUM Toronto
105.3	CFCA Kitchener
105.7	CHRE St Catharines
106.1	CKLA Guelph
106.5	CILQ Toronto
106.9	WYRK Buffalo
107.5	CING Burlington

It is quite a surprise for most people in the UK when they realise what a wide choice of viewing and listening is available to the population across the Atlantic. In order to complete this Chapter I will now list a selection of satellite TV services which are available in the USA to TVRO owners, and, once again, the choice is vast.

Table 8.3 Satellite TV services available in USA and Canada

Title	*Programme type*
The knowledge network	Canadian educational programme
Access Alberta	Canadian educational programme
Premier choix/TVEC	Movies, sports and specials in French. Canada
Superchannel/First choice	Movies, sports and specials in English. Canada
Atlantic satellite network	CTV network programme. Canada
La sette	Programmes from France distributed. Canada
TV Ontario	Educational programme. Canada
TSN (The Sports Network)	Sports programme. Canada
Global television	Daily news, sport and network feeds. Canada
Much music	Rock music videos. Canada
CHCH, Hamilton, Ontario	Variety programming. Canada
WDIV, Detroit, USA	NBC live and taped programmes. USA
WXYZ, Detroit, USA	ABC live and taped programmes. USA
CBC, north-pacific	Canadian variety programmes in English. USA
TCTV (Telemedia Com. TV)	TVA network variety programmes. Canada
CBC, French-eastern	Network programming in French. Canada
Canadian parliament network	Public affairs programmes in French. Canada
CITV, Edmonton, Alberta	Variety Programming in English. Canada
CBC, north-eastern	Network programming in English. Canada
CBC, Montreal	Network programming in English. Canada
WTVS, Detroit, USA	PBS affiliate feeds and variety programming. USA
BCTV, Vancouver	Canadian CTV affiliate feeds and variety programming.
WJBK, Detroit, USA	CBS live/taped programming. USA
Canadian parliament network	Public affairs programmes in English. Canada
Learn Alaska TV network	Educational programme. USA

Table 8.3 *continued*

Title	*Programme type*
Alaska satellite TV project	Variety programming. USA
TNN (The Nashville Network)	Country–variety. USA
WGN, Chicago	Variety programming. USA
The Disney channel	Family entertainment. USA
Showtime	Movies, sports, specials. USA
SIN (Spanish International Network)	Spanish language programming. USA
CNN (Cable News Network)	News coverage and features. USA
ESPN	Professional, college and amateur sports. USA
TMC (The Movie Channel	Movies. USA
CBN (cable network)	Family, christian entertainment. USA
Request TV	Movies on a pay-per-view basis. USA
C-SPAN	US public affairs programme
WOR, New York	Variety programming.USA
PTL satellite	Religious programming. USA
WTBS, Atlanta	Variety and family programming. USA
Cinemax	Movies, sports and specials. USA
Galavision	Spanish language programming. USA
USA cable network	Variety programming. USA
The discovery channel	Science, educational and variety programming. USA
AVN (Alternate View Network)	Educational and religious programming. USA
HBO (Home Box Office)	Movies, sports and specials. USA
Senate republican conference	US political programming. USA
NTU (National Technological University)	Educational programming. USA
AMCEE	Educational programming. USA
Campus satellite network	Variety programming for college campus distribution. USA
ESPN	Professional, college and amateur sports. USA
Showtime	Movies, sports and specials. USA
Biznet satellite cinema	News, variety and movies. USA
Yonkers horse racing	Occasional US horse racing
Radio televisione Italiana	Daily Italian news and sports feeds. USA

Table 8.3 *continued*

Title	*Programme type*
Nickelodeon	Children's programmes. USA
PTL satellite	Interdenominational family programming. USA
TBN (Trinity Broadcast Network)	Family christian programming. USA
FNN (Financial News)/ Score	Live business, financial, sports and events. USA
Viewers choice	First run movies on a pay-per-view basis. USA
SPM (Satellite Program Network)	Movies, entertainment and financial self-help. USA
MTV (Music Television)	Advertiser-supported music videos. USA
EWTN (Eternal Word TV Network)	Catholic spiritual growth and family programmes. USA
Open net	Public affairs programming. USA
VH–1/Video Hits One	Advertiser-supported video music. USA
HTN (Home Theater Network)	Family programming, including movies, specials and travel. USA
TLC (The Learning channel)	Learning programmes and college credit courses for adults. USA
Lifetime	Programming about health and self-development. USA
Reuter monitor service	News and price retrieval in commodity and money markets. USA
NJT (National Jewish TV)	Family jewish programming. USA
BET (Black Entertainment TV)	Black-oriented movies, specials and sports. USA
The weather channel	Live, constantly updated national weather service. USA
MSN (The Information Channel)	Weekly series on business, travel and health. USA
HSN (Home Shopping Network)	Designed for shopping via television. USA
USA Blackout Network	Programming used in place of blacked-out sporting events. USA
Cinemax	Commercial-free movies and specials. USA

Table 8.3 *continued*

Title	*Programme type*
A & E (Arts and Entertainment)	Distinctive entertainment from around the world. USA
Bravo	International films, performing arts and cultural programmes. USA
MSG (Madison Square Garden)	Sport and variety programmes. USA
NCN (National Christian Network)	Multi-denominational religious programmes. USA
AMC (American Movie Classics)	Movies made before 1970. USA
The playboy channel	Adult-oriented game shows, music, specials and films. USA
The heartbeat network	Religious and family programmes. USA
Playboy private ticket	Occasional adult programming. USA
SNL (Success-N-Life Network)	Educational, exercise and variety Christian programmes. USA
The silent network	Programming in sign language and voice. USA
Hit video USA	Contemporary hit music videos. USA
WPIX–New York	Family-oriented programming movies and sport.USA
The nostalgia channel	Movies, comedy and variety programmes. USA
HTS (Home Team Sports)	Sports service from the capitol region. USA
Nightline television network	Religious and variety programming. USA
NASA Contract Channel	Live NASA missions and related events. USA
AFRTS (American Forces Radio and TV)	Independent programming and network feeds. USA
PSN (Private Satellite Network)	Videoconferencing and private video feeds. USA
HSN (Hospital Satellite Network)	Medical programming and uplifting movies. USA
ACTS (American Christian TV)	Family christian entertainment. USA
VMT (Vanderbilt Medical TV)	Medical programming for doctors and hospitals. USA

Table 8.3 *continued*

Title	*Programme type*
BTN (Baptist Telecom. Network)	Religious programming for churches. USA
CMTV (Country Music TV)	Country-oriented videos and programmes. USA
Selec TV	Feature films and adult movies. USA
American extasy	X-rated movies. USA
KTVT–Dallas	Movies, sport and variety programming. USA
The university network	Nonsecular programming with Dr Gene Scott. USA
PASS (Pro Am Sports System)	Sports from Michigan, Ohio, Indiana. USA
Telebet racing	Occasional horse racing. USA

The above list was extracted from an American Publication *Satellite World,* and readers will appreciate that we have a long way to go in Europe before we can hope to match the wide selection of programmes which are available in North America and Canada.

9 Typical programmes on satellite

This Chapter is devoted to listings of typical programmes which are currently available on satellite TV in the UK.

Music box

0600, 1200, 1800 GMT Hear and Now with Julie.
0700, 1300 GMT First Sight repeat
1900 GMT Rox Box from Belgium
0800, 1400, 2000 GMT Sound and Vision with Gary
0900, 1500, 2100 GMT Home Movies with Martin
1000, 1600, 2200 GMT The Eurochart with Gaz and Edwina
1100, 1700, 2300 GMT Sunie Hits the Screen

Lifestyle

0900 GMT It Figures
0930 GMT Sally Jessy Raphael
1000 GMT Farmhouse Kitchen
1030 GMT A Better Read
1100 GMT Good Sex
1155 GMT Take Kerr
1200 GMT It's a Vet's Life
1230 GMT Divorce Court

SAT 1

1500 CET Kimba, der weisse Lowe
1530 CET Lassie
1600 CET Musicbox
1700 CET Raumschiff Enterprise
1800 CET Westlich von Santa Fe
1830 CET APF Blick news
1845 CET Extrablatt
1945 CET Affare Nabob
2130 CET APF Blick news, sport and weather
2215 CET WM-Das Wirtschaftsmagazin
2245 CET JuxBox Extra
2300 CET Matt Houston
2345 CET APF Blick late news

TV5

1900 CET The Village in the Clouds children's programme
1930 CET 30 Million Friends animal magazine
2000 CET Number 1 Claude Nougaro variety show
2130 CET The World is a Stage documentary series
2200 CET News relay of main French network news
2230 CET The Dance Hostess film

Mirrorvision

1600 GMT Boney
1700 GMT Tag the Intrepid Fox

2000 GMT Star Struck
2200 GMT Silence
2330 GMT Oddballs
0100 GMT Games that Lovers Play

The children's channel

0700 GMT Roustabout
0730 GMT Shazam
0800 GMT Inspector Nasy and Laff-a-bits
0830 GMT Cats and Co.
0900 GMT Transformers, Magic and He-Man
1000 GMT Jack in the Box
1100 GMT Roustabout
1400 GMT Jack in the Box

The arts channel

0600 GMT Arts Review Today magazine
0615 GMT Joseph Heller in conversation
0705 GMT Dostoevsky notes from underground
0720 GMT Tolstoy The Kreutzer Sonata
0735 GMT Rafael Quartet Beethoven Grosse Fuge

ATN/Filmnet

1700 CET The Duellists film
1840 CET Tom and Jerry cartoon
1850 CET Diary of a Mad Housewife
2030 CET The Island

2220 CET The Professionals
2310 CET De Mannetjesmaker

Sky Channel

0745, 0815, 0900, 0945, 1045, 1130, 1215 GMT Sky Trax
1315 GMT Skyways drama
1405 GMT New Animal World
1435 GMT Winter World Skiing
1500, 1545, 1603 GMT Sky Trax
1730 GMT The Brady Bunch
1800 GMT The Lucy Show
1830 GMT Green Acres
1900 GMT Mork and Mindy
2000 GMT Police Woman
2055 GMT The Untouchables
2150 GMT NHL Ice Hockey
2255 GMT Sky Trax

Teleclub

1600 CET Follow that Rainbow
1740 CET Trailer
1800 CET Gandhi
2110 CET Three against Three
2300 CET Escape to Victory
0100 CET 100 Days in Palermo

Screen sport

1700 GMT News
1710 GMT Pro Football Arm Wrestling

1730 GMT World Sports
1750 GMT Turbo
1930 GMT News
1940 GMT Americal Basketball
2025 GMT International Table Tennis
2200 GMT News
2210 GMT Take Stocks
2240 GMT World Sports
2255 GMT World Wide Wrestling

Prem1ere

1500 GMT Victoria Regina
1630 GMT Robotechs
1700 GMT Summer Switch
1800 GMT Careful He Might Hear You
1945 GMT Dune
2200 GMT The Osterman Weekend
0000 GMT The Man Who Loved Women
0145 GMT The Evil that Men Do

CNN

1100 CET Daybreak
1130 CET Business Day
1200 CET Daybreak
1400 CET Daywatch
1700 CET Take Two
1900 CET Newsday
2000 CET Headline News
2100 CET Newsday
2200 CET Newswatch
2330 CET Showbiz Today
0000 CET Moneyline
0030 CET Crossfire

0100 CET Prime News
0200 CET Larry King Live
0300 CET CNN Evening News
0400 CET Moneyline
0430 CET Crossfire
0500 CET Newsnight
0600 CET CNN Headline News
0630 CET Newsnight Update
0730 CET Moneyline
0800 CET News Overnight
0900 CET Larry King
1000 CET CNN Headline News
1030 CET Showbiz Today

Europa

1455 CET Sport with live Tennis from Antwerp
1800 CET Programme Schedule
1805 CET Tik-Tak children's programme
1810 CET Bolke the Bear
1820 CET Childrens programme
1850 CET The Wild World of Animals documentary
1915 CET Sports report
2032 CET World Watch news
2115 CET Opera Maria Stuart, Donizetti
2318 CET Agenda
2348 CET World Watch news
2353 CET Weather report
2355 CET Programme schedule

WPN

0800–0930 CET Rebroadcast of previous afternoon's programmes
0937 CET In View of Holland from the Netherlands
1005 CET Nederland

1025 CET Guerre à la Guerre
1200 CET Interval
1400 CET Live from Washington America Today
1430 CET Arts America
1530 CET English American Style
1537 CET Hello Jerusalem magazine from Israel

RTL+

1700 CET Gewinnspiel
1705 CET Soap Opera
1730 CET Magazin
1740 CET Kurznachichten
1758 CET Magazin
1800 CET RTL–Spiel
1815 CET Musik Deutsche Szene
1835 CET Blick in die Welt
1853 CET 7 vor 7-News-Show/Wetter
1918 CET Karlchen
1922 CET RTL–Spiel
1930 CET Knight Rider
2022 CET RTL–Spiel
2030 CET Spielfilm Grosser internationaler Spielfilm
2205 CET RTL-Spiel
2210 CET Wie geht's
2240 CET Wetter
2245 CET Horoskop
2250 CET Betthupferl

The programme information above is based on details published in a monthly magazine called *Satellite TV Europe*, which is currently available on subscription from 21st Century Publishing, 531–533 Kings Road, London SW10 0TZ.

The programme schedules for Moscow 1 and RAI Uno programmes are published in the daily newspapers of Russia and Italy. Copies of these can be obtained in London and the larger cities in the UK. These programmes are not only available on satellite TV but they are exactly as seen by the public in Russia and Italy.

The installation 10

If you decide to have a TVRO installation, the first thing is to decide if you have a suitable location for mounting the dish. As all the satellites are in geostationary orbit above the Equator, it is essential that you have a clear line of sight to the satellites you want to watch. A rough guide as to the suitability of a site, in the UK, is if there is a clear view to the South from an angle of about 20 degrees above the horizontal. If you are prepared to have a dealer install your system, then he will probably do a site survey and advise where it is possible to install the dish. If you wish to carry out your own survey then you probably only require a simple compass and a protractor to determine a suitable location. It is worth bearing in mind that trees can interfere with the reception of satellite signals, so if your survey is in autumn or winter do consider any local trees which will be full of leaves at other times of the year.

Having fixed the dish mount, it is then necessary to equip the dish with its electronics. These electronics vary in their sophistication, from one LNB, mounted very accurately at the prime focus, to perhaps two LNBs with a suitable waveguide coupler, to a single LNB fitted with a polarotor. This is an LNB fitted with an electromechanical membrane to enable the selection of either polarisation.

Polarisation is something you will hear discussed in connection with satellite TV, but because this book has avoided too much technology, the subject has not been mentioned. Suffice it to say that for various technical reasons, not the least being conservation of power in the satellites, and as a means of minimising interference between channels, a system is used whereby some of the transmitted signals are sent out with their wave pattern displaced by 90 degrees. This is called the polarisation of the signals and each channels polarity is identified by either 'X' and 'Y', or horizontal and vertical. It is for this reason that, in order to be able to receive signals from all the channels on a particular satellite, you have to have the facility of rotating one LNB through 90 degrees, or have two LNBs, each set to one of the polarities, or an LNB with a polarotor which can electrically switch from one to the other. The simplest and cheapest system is to have one LNB and then whenever you want to change polarity you go to the dish and just turn the LNB through 90 degrees.

Table 10.1 Look angles

	Azimuth (degrees ETN)	*Elevation (degrees)*
Eutelsat 1–F1 @ 13 degrees East		
London	163.2	29.7
Birmingham	161.7	28.5
Leeds	161.8	27.2
Budapest	186.0	35.7
Tenerife	129.9	43.7
Intelsat VA–F11 @ 27.5 degrees West		
London	213.3	25.7
Birmingham	211.4	25.4
Leeds	210.8	24.4
Budapest	233.7	20.7
Tenerife	202.2	54.8
Gorizont 7 @ 14 degrees West		
London	197.3	29.6
Birmingham	195.4	29.0
Leeds	195.0	27.7
Budapest	219.8	27.8
Tenerife	174.6	56.8
Eutelsat 1–F2 @ 7 degrees East		
London	170.8	30.6
Birmingham	169.0	29.5
Leeds	169.1	28.2
Budapest	194.1	34.9
Tenerife	137.4	48.0
Eutelsat 1–F4 @ 27.5 degrees East		
London	167.0	30.2
Birmingham	165.3	29.0
Leeds	165.4	27.8
Budapest	190.1	35.4
Tenerife	133.5	45.9

Having installed the dish and its electronics, the next job is to run the cable from the dish into the house and connect up to the satellite tuner. The tuner is then connected to the TV set, switched on and we are ready to look for the satellite.

The first thing you need to know is the position of the satellite so that the dish can be pointed. The position of the satellite is known as the 'look angle', and this varies according to where your site is located. Table 10.1 shows the positions of some of the more popular satellites, and how their 'look angles' differ for different locations.

Having decided on your satellite, say Eutelsat 1–F1, and your site is in West Yorkshire. Set the bearing of the dish (azimuth) to approximately 162 degrees ETN (East of True North) and adjust the elevation to 27.5 degrees above horizontal. Tune the satellite tuner to one of the channels on Eutelsat 1–F1, say Teleclub, and then watch the TV set, whilst moving the dish slowly. First move the bearing for about 10 degrees either side of 162 degrees. If a picture, however poor at this stage, is seen, leave the bearing and make similar adjustments to the elevation setting. As the picture reaches an optimum, leave the elevation adjustment and now rotate the LNB to obtain the best picture. This sets the polarisation to the 'Y' position. Continue making slight adjustments to all three positions, azimuth, elevation and polarity until the best picture quality is obtained. If you now make a slight mark to show where the LNB is set, you should be able to rotate it through exactly 90 degrees, return your satellite tuner to the RAI Uno channel and see the pictures from Italy. As a final tuning procedure it is often easier to tune for a 'null' picture by adjusting the LNB for minimum signals on say RAI Uno, whilst the polarity is on the 'Y' setting. Having established the optimum signals on these two channels it is now simply a matter of tuning the satellite tuner to any of the Eutelsat 1–F1 channels, remembering to set the LNB to the appropriate polarity.

So that you can be absolutely sure that the dish is correctly aligned, it is a good idea to 'tune' for an optimum signal. Most of the better tuner modules have facilities to read the signal strength on a meter. A pair of simple wires from the tuner out to the dish will enable a small portable meter to be read whilst actually adjusting the dish position. This is the best way to ensure optimum tuning, although a very acceptable picture can usually be obtained without going to such lengths.

If you decide to have a dish with an az/el mount it will be necessary every time you wish to change satellites to make adjustments in both planes, i.e. in bearing and elevation. Using the look angle tables for your location you simply repeat the positioning of the dish exactly as described earlier. If, however, you decide to install a polar mount for the dish, changing satellites becomes a movement in one plane only. It is for this reason that motor-operated dishes are generally the polar mount type. The initial setting-up of a polar mount is rather more involved than with the az/el type but it is worth the effort if you are wanting to change satellites from time to time, and also if you eventually decide to motorise the system.

If you give some thought to the orbit which all the communications satellites are in, knowing this to be above the Equator, you will soon realise that the elevation angle of your dish will be highest when it is facing true South. The angle will decrease as you track East or West of true South, until you drop to the horizon. In order to set up a polar mount it is, therefore, necessary to have a truly vertical post on which to mount the dish mechanism and you must know where South is.

There is another factor to take into consideration when installing a polar mount. This is called the Declination Offset Angle, which is an adjustment to the angle of the dish which varies depending on the latitude of the site (Table 10.2).

Table 10.2 Declination angles

	Axis inclination (degrees)	*Declination offset (degrees)*
London	52.23	6.79
Birmingham	53.10	6.87
Leeds	54.30	6.97
Budapest	47.79	6.36
Tenerife	28.91	4.14

The practical way of installing a polar mount is to find which satellite is nearest to true South. As an example we will assume the site to be in West Yorkshire and the most southerly bird turns out to be Intelsat V–F2 at 1 degree West. Unfortunately this particular bird does not transmit unscrambled television pictures, and it needs experience to recognize the signals. The D.I.Y. enthusiast will probably need to move the dish to Eutelsat 1–F1 at approximately 162 degrees ETN and observe one of the known channels on that bird. Then swing across to Intelsat VA–F11 at approximately 211 degrees ETN and again observe one of the channels. Then, with the aid of your signal-strength meter, keep making slight adjustments to the aiming angle of the dish so that you get optimum signals from both satellites. Before making all these slight adjustments you must, of course, have set the appropriate declination angle for your location. This must not be altered from the listed setting. It can be quite time-consuming to set up a polar mount, and it may be best left to an experienced installation engineer. Once a polar mount is correctly set-up, however, it is a simple operation to scan the entire Clarke Belt from horizon to horizon, and be able to 'view' any satellite you wish that falls between the two extremities. Future satellites for Europe will all be within this range, so once set up the entire group of satellites is at your disposal, both current and future.

Having obtained a satisfactory installation of the dish, the next job is to protect the equipment from the ravages of the weather. The connectors on the cables are particularly vulnerable, and it is essential to prevent the ingress of moisture to the connectors. A self-vulcanising rubber tape is advisable wrapped round the connectors, and for maximum protection this should be topped off with a water-repellant tape such as Denso. Normal PVC type insulation tape is not really suitable as this tends to harden, the adhesives oozes out and it does not prevent moisture getting into the connector.

Once the dish is installed satisfactorily, the next job is to set up the tuner unit. There are more or less only two types of tuner. One has a dial, and possibly an upper/lower band switch. The other type has a progammable memory and an LED channel indicator.

With the dial type you simply turn the knob and, if necessary, note the position of the stations. In the case of the programmable memory you simply input each channel as it is found. From then on it is only necessary to recall any channel from the tuner memory whenever you wish to view. As an example, overleaf is a table of the channel numbers which apply to the SATVRN TDM1200 series of tuners for four different satellites.

As this particular tuner has an infra-red remote controller it is possible to sit in your armchair and change the programme channels, high/low band switch and polarisation, together with other refinements such as a clamp switch and different sound systems. This is quite a sophisticated domestic tuner, and if it is coupled with a remotely-operated motor drive for the dish, you have a very comprehensive television programme selection at your fingertips. As an added bonus the LED display operates as an accurate clock when not being used to display the selected channel number.

That just about sums it up as far as installation is concerned. If the prospects of DIY are too daunting, you should seek an experienced satellite TV installer to fix up the system. If, on the other hand, you feel confident to do it yourself, then have a go, you will become more and more interested as you progress. Several of the more reputable manufacturers will be only too pleased to offer help and guidance. This is particularly so at the present time when experienced installation companies are rather thin on the ground and, unfortunately, there are a few disreputable installers who are taking

Channel	*Programme*	*Satellite*	*Band*	*Polarisation*
07	Teleclub	1	L	Y
11	RAI Uno	1	L	X
11	TV5	1	H	X
11	New World Channel	1	H	X
11	SVT–1	2	H	Y
13	Prem1ere/Children's Channel	4	L	X
17	SAT–1	1	H	Y
20	World Net	1	H	X
21	3–SAT	1	L	X
27	RTL+	1	l	Y
37	Screen Sport/Lifestyle	4	L	X
38	ATN/Filmnet/WPN	1	L	Y
41	Cable News Network	4	L	Y
44	SVT–2	2	H	Y
45	Europa	1	L	X
45	NRK	2	H	X
45	Mirrorvision	4	L	X
47	Sky Channel	1	H	X
52	Music Box	1	H	Y

Satellite 1 : ECS (Eutelsat–1) F1 @ 13 degrees East
Satellite 2 : ECS (Eutelsat–1) F2 @ 7 degrees East
Satellite 3 : ECS (Eutelsat–1) F4 @ 10 degrees East (new channels soon)
Satellite 4 : Intelsat VA F11 @ 27.5 degrees West
Polarisation X = ECS X or Intelsat H
Polarisation Y = ECS Y or Intelsat V

advantage of some of the public who are rather gullible and ignorant about satellite TV. As a guide you should receive *better quality* pictures from satellite than you normally get from terrestrial television stations. The main reason for this is the fact that satellite TV does not suffer from 'ghosting' and similar problems because it is not affected by reflections from hills and large buildings. If you see satellite TV which is inferior to terrestrial TV the chances are that the installation has not been done correctly, or the receiving equipment is of an inferior quality.

Satellite types and identification 11

Communication satellites fall into many categories. The majority used for television transmission are owned by private companies, like Intelsat, the International telecommunications satellite organisation. Services on the various satellites are leased to organisations, such as the programme companies for television and radio. Telephone companies lease channels on the satellites for telephone links and data links, etc.

As equipment develops, more sophisticated versions are brought into use. As an example, let us examine the Intelsat series of satellites. The very first 'Early Bird' became Intelsat 1. Since then each type of Intelsat has been given a series, indentified by Roman numerals. Series 1, 2 and 3 are no longer operational, but there are now several series (IV, IVA, V, VA and VI) Intelsat satellites in operation.

Each satellite is given a flight number, and this is a suffix added to the type number. Confusion sometimes arises in magazines when only the type number is referred to. In order to positively identify an Intelsat satellite, not only is it necessary to know its type but you also need to know the flight number.

The position of each satellite in the Clarke Belt is published by the reliable magazines, and this position is important when establishing which satellite you wish to tune in to.

The position of the satellites can be changed from ground control. This is used to replace ageing satellites with new launches, so you might see a change of type in a particular location from time to time.

Here are a few examples of Intelsat 'birds'

Intelsat IV – F8 located 1.0W (replaced with V – F2)
Intelsat IVA – F1 located 31.0W
Intelsat IVA – F2 located 57.0E
Intelsat IVA – F4 located 21.5W
Intelsat V – F1 located 57.0E (replaced with VA – F12)
Intelsat V – F2 located 34.5W (replaced with V – F4)

Intelsat V – F3 located 24.5W (replaced with VA – F10)
Intelsat V – F4 located 27.5W (replaced with VA – F11)
Intelsat V – F5 located 63.0E
Intelsat V – F6 located 18.5W
Intelsat V – F7 located 60.0E (replaced with VA – F12)
Intelsat V – F8 located 53.0W
Intelsat VA – F10 located 24.5W
Intelsat VA – F11 located 27.5W (currently carries Prem1ere, Mirrorvision, Screen sport, CNN, etc.)
Intelsat VA – F12 located 60.0E

The other groups of satellites in which Europeans are most interested are the ones operated by Eutelsat (The European Telecommunications Satellite Organisation), sometimes referred to as ECS (European Communications Satellites), and the Russian ones. Here is a list relating to these.

Eutelsat 1 – F1 located 13.0E (currently carries RAI Uno, ATN, Teleclub, Sat 1, Music box, etc.)
Eutelsat 1 – F2 located 7.0E
Gorizont 7 located 14.0W
Gorizont 9 located 53.0E
Gorizont 10 located 80.0E

In reading about satellite TV you will probably come across references to *beams*. These refer to the size of the 'footprint' from each transponder, or set of transponders. A general rule of thumb is that the larger the footprint the weaker is the signal reaching the ground. Conversely, the smaller the footprint the signal is stronger, and it follows from this that a smaller dish is possible for a smaller 'footprint'.

A global beam indicates that a large area of the Earth's surface (up to 40% is covered. Large dishes are needed generally to receive global beam transmissions, and these will normally be outside the scope of the readers of this book.

A hemisphere beam covers up to 20% of the Earth's surface, and somewhat smaller dishes are adequate.

Spot beams, on the other hand, are the ones which only require small (1.2 metre to 1.8 metre dishes, and these are the type of beams which are used for most of the programme services we have referred to in this book.

Future developments 12

The future of satellite television is likely to be very interesting indeed. Present technology is already being improved and will no doubt be responsible for many of the developments listed in this Chapter. As technology improves then so will the facilities available on satellite.

One of the changes, already being discussed by the owners of the Europa channel, is the transmission of sub-titles in various languages to accompany the multi-lingual programmes.

In a similar vein, some companies are already considering transmitting several different sound programmes to accompany the pictures. Although this is not as simple as sub-titling it is probably a more acceptable solution for most viewers. By using such a system it would be possible to have the sound which accompanies the pictures in several different languages, so that a programme or film with English dialogue could be accompanied with simultaneous translations into, say, Dutch, German, French and Italian. A system like this would not only make the programmes more enjoyable, it would be most useful for people wishing to study languages. A simple switch on the tuner unit would enable the selection of the desired sound channel.

A few of the satellite channels, Music Box is one, are already transmitting stereophonic sound to accompany their pictures. It is likely that more and more channels will adopt this as time goes on.

New satellites are scheduled to be launched over the next few years. Some of these are likely to have higher power transponders than the ones we have been dealing with so far, so it will be possible to receive them with smaller dishes than the ones at present.

The higher power satellites will transmit a system called DBS which is designed to be received by viewers direct into their own homes. It is quite likely that the IBA and BBC will be using these satellites for transmission of their programmes. It is also possible that other programme providers will use these higher powered-satellites to transmit pay-per-view first run movies, again direct into homes.

Although the technology exists today it is likely to be well into the 1990s before a system gets under way.

Other proposals, which may well come to fruition when DBS satellites are launched, include a different colour standard, MAC, which it is claimed has several improvements on the present systems. It is also likely that there will be improved sound quality, including stereo sound, a different aspect ratio so that the TV pictures resemble wide-screen cinema standards, and an improvement in the picture definition, so that larger screens may be used. Other possibilities include several sound channels to accompany each TV programme. This would be particularly appealing in Europe where it would be possible to watch the same pictures with the sound in your own language.

The advancement of electronics is also a probable reason why smaller dishes may become available in the future.

The other possibility in the not-too-distant future is to have a computer chip built into the satellite tuner. Such a device would enable the viewer to select the programme he wanted to watch, then the tuner would arrange that the dish would be turned and lined up on the appropriate satellite. It would select the correct polarisation and see that the correct sound system was switched into use. It would also be possible to pre-programme the system, in a way similar to some of the more expensive video tape recorders, so that certain programmes would automatically be selected at different pre-selected times of the day. The technology for these systems is already in existence so it is probably only a short time before somebody comes up with such a turner.

An exciting new future is about to open up before us in the world of satellite TV. I hope the information in this book has whetted your appetite, and that you too will soon be enjoying a new experience with television from space extending your viewing choice way beyond the customary three or four terrestrial channels.

The world of satellite TV is a very fluid one, and changes are taking place all the time. There have been changes in the few months since the preparation of this book was started. Worldnet is no longer shown on the Filmnet channel, Mirrorvision has ceased, and there is a possibility that RAI Uno will soon move to another satellite and its slot on Eutelsat 1 is likely to be taken by a programme originating in Luxembourg, called RTL+. ATN/Filmnet is now scrambled, and Intelsat VA–F12 at 60° East, over the Indian Ocean, is transmitting four German programmes: WDR–3; Music Box; BR–3 and Eins Plus.

Plans are already well advanced for a higher-powered satellite, devoted entirely to television, to be operational by the spring of 1987. This is to be called the Astra project, and it will provide 16 channels of television programmes specifically for Europe. It is hoped that some of the programmes already carried on Eutelsat 1–F1 and Intelsat VA–F11 will be transferred to this bird.

The final item concerning future developments is prices. Already, in the past twelve months or so, prices have come down considerably. The £1000 barrier for a simple domestic system has been breached, and people in the industry are already striving to break the £500 barrier before the end of 1987.

Glossary

Ariane The European expendable launch vehicle (ELV), used to put communications satellites into orbit.

Az/el mount This is a type of mounting for a dish which has separate movements for azimuth and elevation. It is simpler to set up than a polar mount, but requires more adjustments if you change satellites.

Azimuth The bearing, relative to positions on the compass. North being 0 or 360 degrees, East is either 90 degrees East or 270 degrees West of North. South is 180 degrees East or West, and West is either 270 degrees East or 90 degrees West of North.

Bird A slang term, of American origin, which means a communications satellite.

Clarke belt The name given to the orbit in which all the geostationary communications satellites are positioned. It is directly over the Equator, about 23,000 miles out from Earth.

DBS The initials stand for 'Direct Broadcasting by Satellite', and refer to the system which it is hoped will eventually replace earth-based transmitting stations and provide domestic television from satellites direct into the home.

Dish The common name for the parabolic reflector which collects the weak signals from the satellites and focusses the energy to the collector on the first item of electronics in the receiving chain, the LNB.

Earth station This is the complex, usually owned by Telecommunications Authorities, such as British Telecom in the UK, to where the signals from the studios are sent, which are then 'uplinked' to a particular satellite, ready for retransmitting back to Earth by that satellite. Earth stations use very large dishes for this purpose.

ECS The initials stand for European Communications Satellite.

ETN These initials stand for East of True North, and are used to indicate the azimuth bearings. It is more common to use these bearings in listings of satellite positions than to refer to positions which are West of North.

Elevation This is the angle of tilt required on a dish in order to 'see' a particular satellite. Horizontal is 0 degrees and vertical is 90 degrees.

Eutelsat European Telecommunications Satellite Organisation.

Flight Satellites often carry the name of their owner organisation, such as Intelsat or Eutelsat. In order to distinguish different satellites they are given flight numbers.

Footprint This is an area on Earth where the signals from the satellites are beamed. Maps showing footprints are published to show the signal strengths of different satellites in various locations.

Geostationary Used to describe the satellites which are located in the Clarke Belt, and which rotate round the Earth at the same speed as the Earth's rotation, so they appear stationary when viewed from Earth.

Hertz or Hz The unit of frequency. It is equivalent to cycles per second.

Intelsat The name of one of the American companies who own quite a large number of communications satellites.

LED Light Emitting Diode. An electronic lamp, used as an indicator. The main advantage being its very long life and the small amount of electric current it uses.

LNA Low Noise Amplifier. One of the electronic devices which is used at the prime focus of a dish to amplify the weak satellite signals. The frequency down-converter is a separate item.

LNB Low Noise Block Down Converter. An advanced version of the LNA, which is nowadays the most commonly used device at the dish prime-focus point. This, as its name suggests, incorporates the frequency down-converter in the same box. it converts a block of intermediate frequencies which can then be fed to several satellite receivers, which can be tuned to different channels.

LNC This is simply an LNA and a frequency down-converter, assembled in one box.

Look angle This is the angle of elevation for a particular satellite from a particular location.

MAC Multiplexed Analogue Component. A new colour television standard, which overcomes several of the problems associated with current systems like PAL and SECAM.

NTSC National Television System Committee. The first practical broadcast colour television system, developed in the USA, and still the system which is used in the Americas.

Orbit The path described by a satellite. In the case of most communications satellites this is geostationary, although some satellites are in geosychronous orbits and these have to be 'tracked' by the Earth dishes as they appear to move across the sky.

PAL Phase Alternation Line system is a variant on NTSC, which was developed to reduce the effects of certain types of distortion which can occur with NTSC. It is the most widely adopted system which is used in Europe and various other countries.

Parabola This is the shape of the profile of the majority of dishes used for the reception of satellite signals. The more accurate the profile is determines the efficiency of the dish.

Polar mount This is a mounting for the dish, where the axis of said dish is lined up with the Earth's axis, and thus only movement in one plane is necessary to move the dish from one satellite to another.

SMATV Small Master Antenna Television. This is the system whereby one dish, say on an hotel, or holiday complex, is used to feed a number of separate satellite tuner units.

Scalar horn A device which is fitted to the front end of most LNBs to help concentrate the received signals and focus them accurately on to the LNB.

SECAM Sequence à Memoire is the third system of colour television, yet another variant on the original NTSC system, which has been adopted as the standard for France and the USSR, in particular.

Shuttle The manned re-usable space vehicle favoured by the Americans for launching satellites into orbit.

TVRO Television Receive Only Terminal. This is the name of installations for the reception of satellite TV into private homes.

Transmitter A transmitter is used by the broadcasting authorities to send the radio-frequency signals out into the ether so that they can be picked up by aerials on people's houses. Transmitters are also used at the Earth stations to send the signals up to the satellites.

Transponder A device used in communications satellites to convert the received signals from a ground station into other signals which are simultaneously retransmitted to Earth for reception by the many dishes which are tuned to that particular satellite channel.

Index